USING ELECTRONIC TESTERS FOR AUTOMOTIVE TUNE-UP

By Albert Wanninger

TAB BOOKS
Blue Ridge Summit, Pa. 17214

FIRST EDITION

FIRST PRINTING—JUNE 1972

Printed in the United States
of America

Hardbound Edition: International Standard Book No: 0-8306-2604-2

Paperbound Edition: International Standard Book No. 0-8306-1604-7

Library of Congress Card Number: 72-80558

Preface

This book puts the spotlight on engine tune-up testers. Its aim is to acquaint the reader with the various kinds of automotive testers that are in common use by both the amateur and full-time professional mechanic.

Each type of tune-up tester is treated in turn by first introducing examples of typical instruments. Next, the scales and controls on these instruments are discussed. Then the method and procedure of connecting the tester and performing a test is explained. Finally, the test results and interpretation are reviewed in detail. Specific instructions on making repairs and adjustments on particular engines are purposely omitted so that full attention will remain on the tune-up testers.

As each chapter deals with a specific piece of test equipment, it is not necessary to read this book in the order in which it is presented. The author invites the reader to skip about the text from chapter to chapter as suits him.

The writer wishes to express his sincere appreciation to H.H. Weston and Edward Pohlmann for their suggestions of material to be included. The author is also indebted to the following manufacturers who have allowed their illustrations to be used in this volume: Peerless Instrument Company, Sun Electric Company, Marquette Corporation, Kal-Equip Company, Fox Valley Instrument Company, Autoscan Incorporated, King Electronics and Manufacturing Corporation, Allen Electric and Equipment Company, A-C Spark Plug Division, Simpson Electric Company, Pioneer Electric & Research Company, and Sears, Roebuck and Company.

It is hoped that the reader will find this volume both informative and enjoyable.

Albert Wanninger

Contents

Chapter 1

Voltmeter and Ohmmeter Tests

The basic test instruments used for diagnosing trouble in an electrical circuit are the voltmeter, ohmmeter, and ammeter. To become proficient in servicing vehicles, the mechanic must develop an understanding of these testers.

An automotive voltmeter is usually combined with another instrument. For example, Fig. 1-1 illustrates a typical voltmeter, which, in this instance, is combined with an ammeter. Another frequent combination is a voltmeter with an ohmmeter.

This chapter will concentrate on uses for the volt-ohmmeter. The chapters on battery-starter testers and charging system analyzers will discuss the ammeter and procedures for using the voltmeter and ammeter together in performing tests.

METER MOVEMENT AND SCALES

The D'Arsonval galvanometer is the type of meter movement used in virtually all electrical and electronic test equipment. For example, the voltmeter, ammeter, ohmmeter, dwell meter, and combustion analyzer all use the D'Arsonval galvanometer. Depending on the particular kind of instrument, the electrical current galvanometer is modified considerably by resistors, capacitors, and special circuits, in order that the type of indication desired will be obtained.

The D'Arsonval galvanometer consists essentially of a movable coil of wire suspended in the magnetic field of a permanent magnet, as shown in Fig. 1-2. Current flowing through the movable coil reacts with the permanent magnetic field, causing the coil to turn against a slight spring tension. A pointer attached to the coil moves across a scale which can be read when the pointer comes to rest.

VOLTMETER

Since the moving coil is very small and sensitive, a voltmeter is constructed by adding resistance in series with

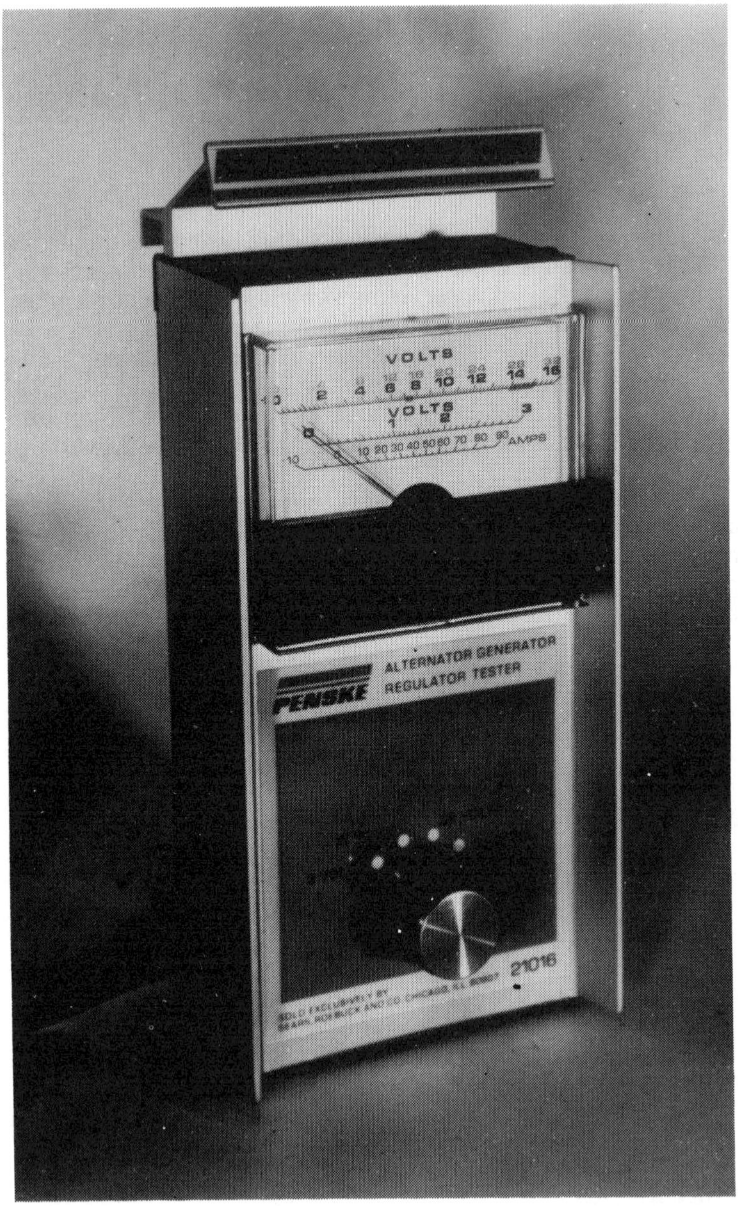

Fig. 1-1. Typical voltmeter combined with ammeter for automotive use.

the coil. This resistance is mounted inside the instrument case.

The voltmeter resistance is frequently printed on the face of the meter, and is usually specified in ohms-per-volt. If the voltmeter resistance is 1,000 ohms-per-volt, and the 16-volt scale on the meter is being used, the voltmeter internal resistance is 1,000 x 16 equals 16,000 ohms. Voltmeters having a choice of scales have a selector knob to insert different resistor values (called multipliers) in series with the moving coil.

Most voltmeters used for automotive work have scales which indicate more than one range of voltage, such as 0-3, 0-8, 0-16, 0-20 and 0-32 volts. A typical voltmeter scale is shown in Fig. 1-3.

Dependable voltmeters have a high standard of accuracy. Commercial instruments used for automotive work generally are guaranteed to be accurate to within 2 percent of their full-scale reading.

The terms **voltage** and **voltage drop** really mean the same thing and refer to the reading you get on a voltmeter when it is properly connected. **Voltage** frequency refers to the total voltage difference between the two sides of a circuit, such as the total battery or total alternator voltage; while **voltage drop** is often mentioned as the drop in voltage in part of a circuit, such as the voltage drop across the contacts of a switch.

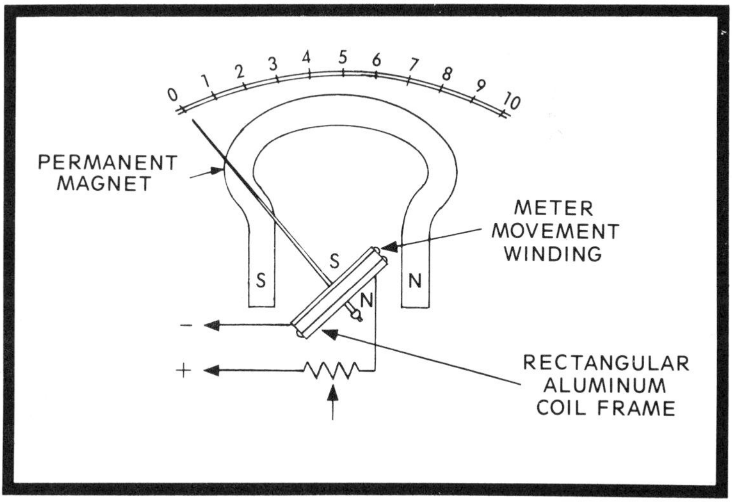

Fig. 1-2. D'Arsonval galvanometer movement with internal resistance for use as voltmeter.

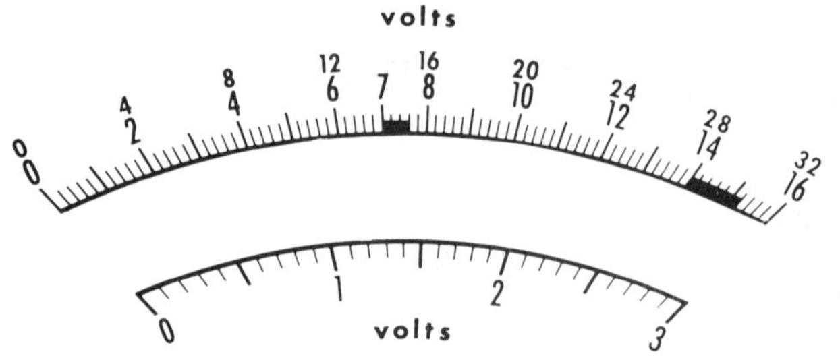

Fig. 1-3. Typical voltmeter scales.

Connecting a Voltmeter

The voltmeter is one of the easiest instruments to use. All you have to do is to connect the two leads from the meter to the two points in the circuit where you want to measure the voltage drop. It is general practice to make the positive lead of the voltmeter red and the negative lead black. If you connect the meter loads to two points in a circuit in which current is flowing, and the pointer reads down-scale, it is merely necessary to reverse the leads. Just touching the leads to any exposed metal ground and an exposed **hot** or **live** metal contact screw or terminal will give a meter reading.

VOLTMETER APPLICATIONS

The voltmeter is an extremely versatile instrument and can be used to test the following components:

- Battery
- Starter
- Relays
- Wire and Cables
- Switches
- Primary Resistor
- Diodes
- Fuses
- Horns
- Lights

This list is by no means exhaustive. It is meant merely to illustrate the broad applicability of this instrument.

Several examples which follow will show the use of the voltmeter. These examples should serve as a guide for other voltmeter applications. Subsequent chapters will illustrate the use of the voltmeter in testing batteries, starters, diodes, and charging systems.

Battery Polarity

Although the positive battery post is larger than the negative post, occasions may arise when a quick test of battery polarity will be useful. To determine the polarity of a battery, switch the selector on the voltmeter to 16 volts and connect the test leads to the battery posts, as shown in Fig. 1-4. If the pointer moves backwards on the scale, reverse the leads. When the pointer moves in the right direction, the plus voltmeter lead is on the plus or positive post of the battery.

Fuses

Automotive fuses can be tested in the circuit with a voltmeter. The voltmeter is set to the 16-volt range and the ignition switch and the switch for the accessory which the fuse protects are both turned on. The negative voltmeter lead is connected to a good ground. The positive voltmeter lead is touched in turn to both sides of the fuse, see Fig. 1-5.

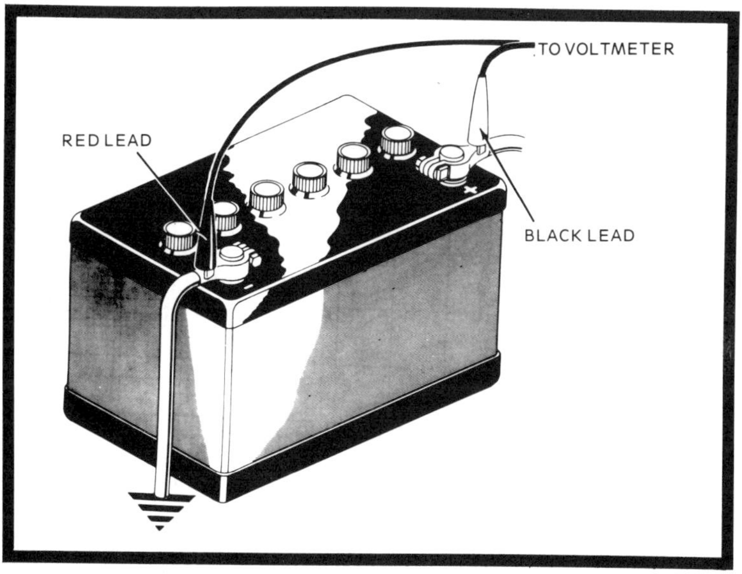

Fig. 1-4. Voltmeter connections for battery polarity test.

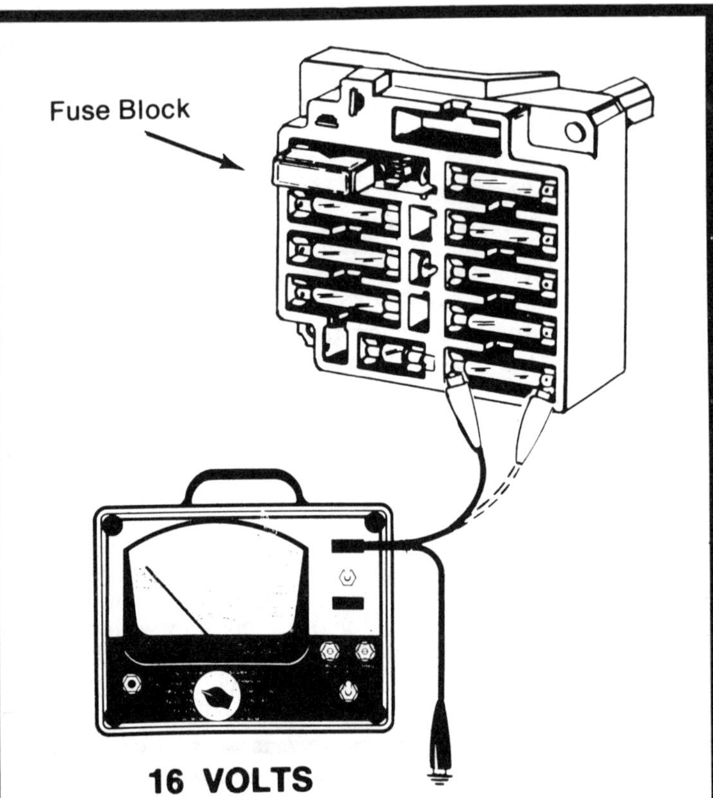

Fig. 1-5. Voltmeter connections for fuse test.

If battery voltage is read on both sides of the fuse, the fuse is good. If, on the other hand, battery voltage appears on one side of the fuse and zero voltage is read on the other side, the fuse is blown. A zero voltage reading on both sides of the fuse indicates an open circuit between the battery and the fuse.

Horns

The horns can be tested with a voltmeter set to the 16-volt range. The negative voltmeter lead is connected to ground, and the positive voltmeter lead is connected to the H terminal of the horn relay (Fig. 1-6).

Operate the horn ring or button and read the 16-volt scale on the meter. The voltmeter should read between 10 volts and battery voltage for a 12-volt system (5 volts and battery voltage for a 6-volt system). A zero reading indicates a defective horn relay, horn button, or a break in wiring between the horn button and relay. If the reading is approximately battery voltage, there is a break in the wiring between the relay and the horn, or the horn is not grounded. If the reading is less than 10 volts on a 12-volt system (5 volts on a 6-volt system) the coil in the horn may be shorted or the relay points may not be making good contact.

No-Start Test

If the complaint is in the no-start category, the voltmeter can be used to determine if the ignition primary is functioning by connecting a voltmeter, set to the 16-volt range, between the negative side of the coil and ground. The engine is then bumped through the distributor points **open** and **closed** intervals. A battery voltage indication during points **open** and a near zero indication during points **closed** interval indicates good primary circuit action.

Primary Resistor Test

This test checks the effectiveness of the primary ballast resistor in limiting current flow across the distributor contacts. The voltmeter is set to the 16-volt range and the positive voltmeter lead is connected to the positive battery post. The negative voltmeter lead is connected to the plus side (battery terminal) of the ignition coil. A jumper wire is connected from the negative side of the coil to ground (Fig. 1-7).

With the voltmeter connected as described, turn the ignition switch on and read the voltmeter. If the meter reading is between 4 and 8 volts the primary resistance is good. If the indication is below 4 volts, the primary resistor has too low a

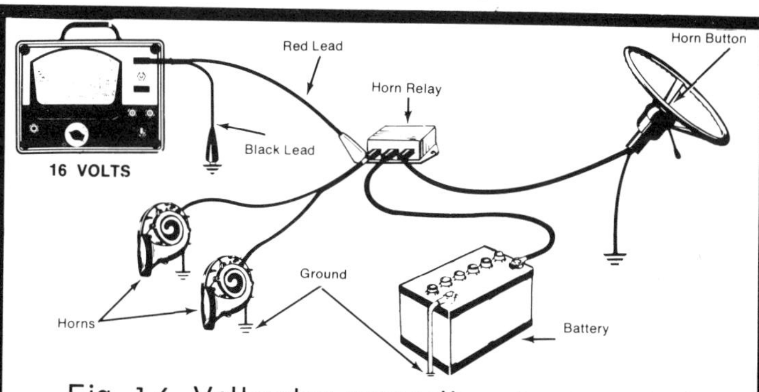

Fig. 1-6. Voltmeter connections for horn test.

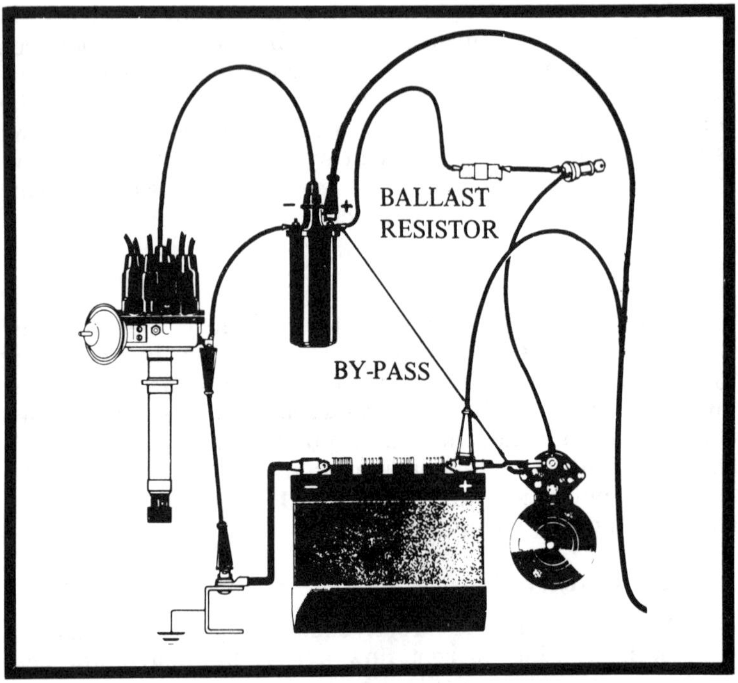

Fig. 1-7. Voltmeter connections for primary resistor test.

resistance. This allows excessive current to flow in the primary circuit, resulting in point burning. If the indication is above 8 volts, the primary resistor has too high a resistance value. In some cases the wiring or the ignition switch may be defective.

If the reading is zero or near zero, the primary resistor is shorted or the primary resistor is located within the coil. The ohmmeter can be used to determine whether the primary resistor is within the coil. Coils with internal primary resistance will measure 3 to 5 ohms. A coil used with an external primary resistance will read 1 to 2.5 ohms. The external primary resistance is approximately 1.5 ohms.

Ignition Switch Test

The primary resistor is usually shorted (bypassed) by the ignition switch during cranking to permit additional energy to reach the coil for better starting. The ignition switch test checks the ability of the ignition switch to short out the primary resistor during cranking.

The lead connections for this test are identical to that described for the primary resistor test, Fig. 1-7, and these two tests are usually performed one after the other.

The ignition switch test is performed by cranking the engine while observing the voltmeter. The jumper wire connected from the coil to ground will prevent the engine from starting.

A satisfactory voltage reading is one volt or less. A voltage reading of more than one volt indicates excessive resistance in the bypass circuit due to faulty ignition switch contacts, faulty bypass connecting wiring, or defective insulated circuit wiring to the ignition switch. The faulty component may be isolated by placing the test leads across the ignition switch and each connection and wire in turn. Readings across each connection should be zero, each wire should not exceed 0.2 volts and switch contacts should not exceed 0.3 volts.

Distributor Resistance Test

Excessive resistance within the distributor can seriously impair an ignition system. A voltmeter can be used to test for excessive resistance because as the resistance increases the voltage drop correspondingly increases.

With the points closed and with the voltmeter set to its lowest range, connect the voltmeter in turn to the following points (Fig. 1-8):

1. The distributor primary terminal of coil to movable arm of breaker points.
2. The movable arm of breaker points to breaker plate.
3. The breaker plate to distributor housing.
4. The distributor housing to engine block.

The voltage drop in either 1, 3, or 4 above should not be higher than 0.05 volts. The voltage drop in 2 should not be higher than 0.2 volts. A voltage drop higher than indicated should be corrected.

OHMMETER

Like the voltmeter, the ohmmeter is also a very versatile instrument (Fig. 1-9). Some of the uses of an ohmmeter are as follows:

1. Checking circuit continuity and locating broken wires or poor connections in such circuits as horns, overdrive, turn signals, lights, accessories.
2. Bench testing of relays, solenoids, and switches.
3. Checking alternator circuit continuity and field coil resistance.

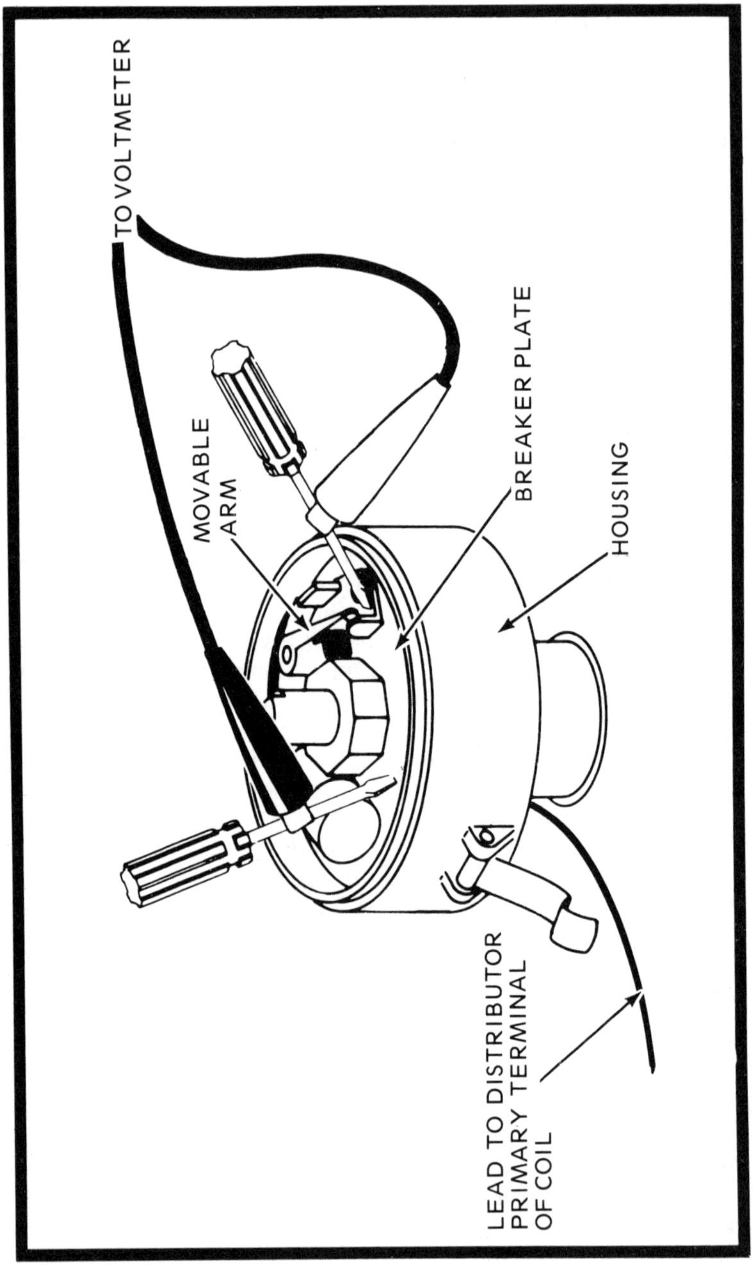

Fig. 1-8. Voltmeter connections for distributor resistance test.

4. Checking ignition system suppressor resistors, distributor caps, rotors, and spark plug wires.

Not every test that can be made with an ohmmeter will be explained in this chapter, but enough illustrations will be given to provide an understanding of how the ohmmeter can be used for the other applications.

Ohmmeter Circuit

An ohmmeter circuit, consisting of a D'Arsonval meter movement connected in series with a battery and a resistor, is

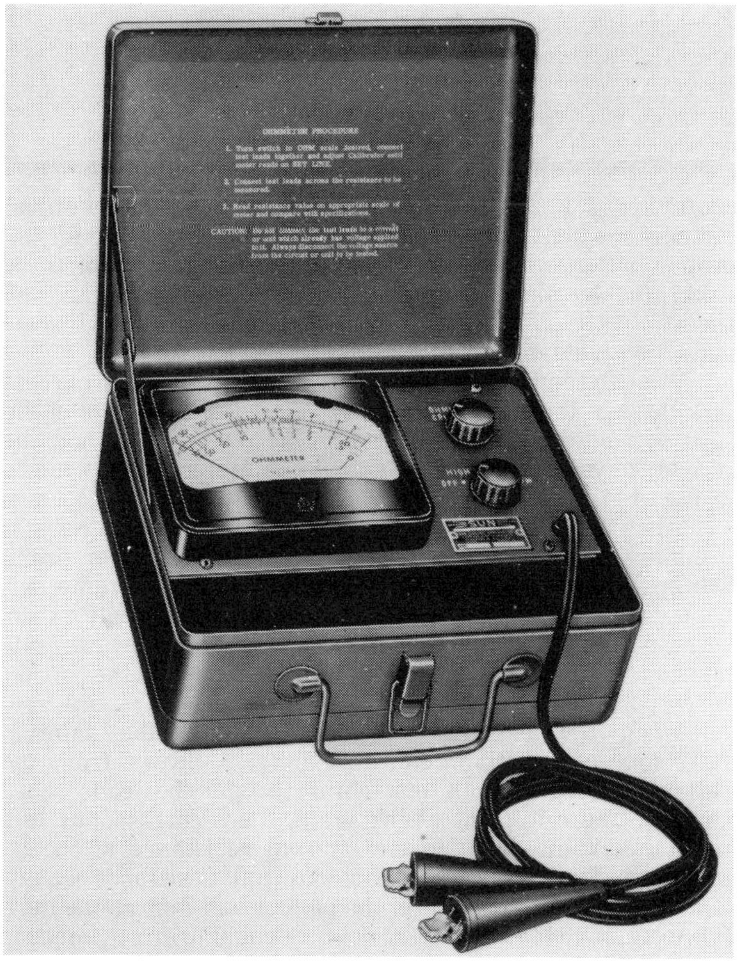

Fig. 1-9. Typical ohmmeter used for automotive work.

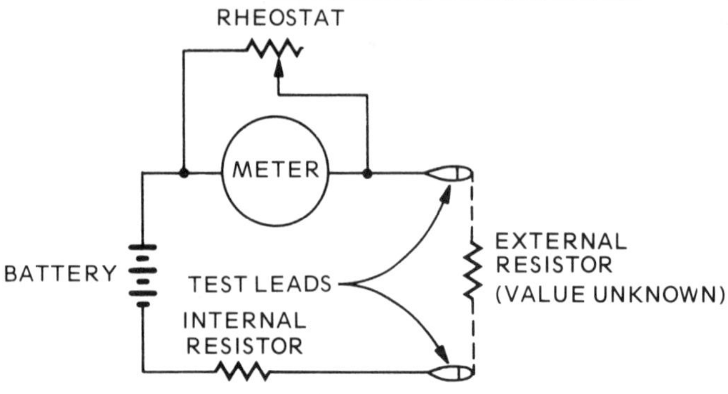

Fig. 1-10. Ohmmeter circuit.

shown in Fig. 1-10. A rheostat is connected in parallel with the meter movement to provide a means of compensating for the drop in battery voltage as the battery ages. The ohmmeter is calibrated by short-circuiting the leads and adjusting the rheostat until full-scale deflection is obtained. Thus, full-scale deflection indicates zero resistance.

When the ohmmeter is connected to a resistor, the current flow through the meter is directly related to the value of the resistor, and the scale is calibrated accordingly to show the resistance value in ohms. A typical ohmmeter scale is shown in Fig. 1-11.

When the leads are open, the value of the external resistor is infinite. This means that the current in the ohmmeter circuit will be zero, and the pointer will remain at the extreme left-hand side of the meter. This point on the scale, then, corresponds to infinite external resistance; while the scale division on the extreme right-hand side of the meter corresponds to zero external resistance.

When the external resistance equals the internal resistance of the ohmmeter, the current drawn from the battery will be one-half that obtained with the leads short-circuits. The mid-point of the scale, then, corresponds to a resistance value equal to the internal resistance of the instrument. Similarly, when the external resistance equals twice the internal resistance, the pointer will rest at one-third full scale. With the external resistance equal to three times the internal resistance, the pointer will rest at one-fourth full scale. It is obvious from the above that the number of ohms per degree of pointer deflection increases steadily as the pointer moves from right to left. This accounts for the nonlinearily of the ohmmeter scale.

Ohm's Law

This law was first formulated by Georg Simon Ohm, a German physicist, in 1826, as the result of many experiments and measurements. His name was first given to the practical unit of electrical resistance, the ohm, in 1861.

The law states that the electron current flow through any conductor equals the potential difference divided by the resistance. It is usually written:

$$I = \frac{E}{R} \text{ or } E = IR$$

where E (volts) denotes the potential difference over the circuit through which the current I (amperes) is flowing. The factor R (ohms) is called the **resistance** of the conductor, and its value depends only upon the dimensions and material of the wire and its temperature. It is entirely independent of E and I.

Using Ohm's Law, an unknown quantity is determined as follows:

To find amperes—divide the voltage by the resistance
To find voltage—multiply the amperes by the resistance
To find resistance—divide the voltage by the amperage

As an example of how Ohm's Law can be used, assume that a starting motor of an automobile requires 175 amperes before rotation of the engine begins, and the resistance of the motor is 0.025 ohm. What voltage must be applied to cause this value of current to flow? Solution:

$$E = IR = 175 \times 0.025 = 4.375 \text{ volts}$$

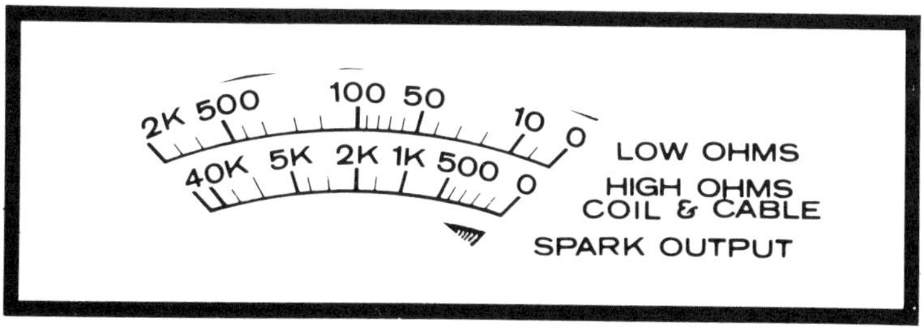

Fig. 1-11. Typical ohmmeter scale with low and high ohms range.

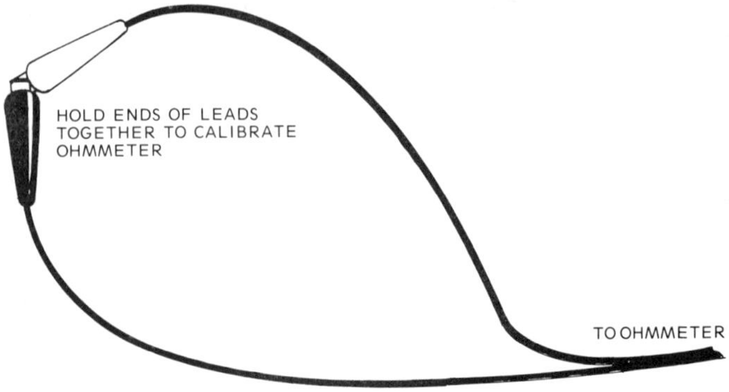

Fig. 1-12. Calibration of an ohmmeter.

Ordinarily, it will not be necessary to compute electrical values using Ohm's Law when performing a tune-up. It is advisable, however, to have a basic understanding of the Law so that meter indications will have more meaning.

Ohmmeter Scales

Ohmmeter ranges vary from low ohms to high ohms. Some of the more common center scale values used on automotive ohmmeters are as follows: 5, 50, 500, 1000, 2000 and 5000 ohms. Fig. 1-11 illustrates a commonly used ohmmeter scale with a low and a high ohmmeter range.

Calibration and Hook-Up

An important caution to observe when using an ohmmeter is never connect it to an external source of voltage, as the meter movement may be permanently damaged. This means that an ohmmeter should never be used when current is flowing through the resistance to be measured.

Another important point to remember is that one side of the circuit or component under test should always be disconnected from its circuitry before the ohmmeter is connected. This is done to prevent the possibility of an erroneous reading due to other resistances in parallel with the resistance under test.

The ohmmeter is calibrated, as was explained earlier, by clipping the two ohmmeter leads together and adjusting the pointer to zero ohms with the set control (Fig. 1-12). Some ohmmeters have a set switch that eliminates the need to clip the leads together. This allows the ohmmeter to be calibrated, for most practical purposes, regardless of whether it is connected to a circuit or not.

After the ohmmeter is calibrated, it is connected across the component to be measured, see Fig. 1-13, and a reading is obtained. The following illustrations will show how the ohmmeter can be used for test purposes.

Spark Plug Wire Resistance

Spark plug wires can be measured with an ohmmeter in the high ohms position. The ohmmeter leads are connected to the two metal end terminals on the wire as shown in Fig. 1-14. The actual resistance reading will vary widely depending on the type of spark plug wire and length. The table in Fig. 1-15 shows typical values for suppressor spark plug wire. No resistance reading (no pointer movement) indicates an open wire which should be repaired or replaced.

Coil Secondary Resistance Test

The secondary resistance of a coil is measured with the ohmmeter in the high ohms position. The distributor wire from the coil is disconnected, and one ohmmeter lead is clipped to

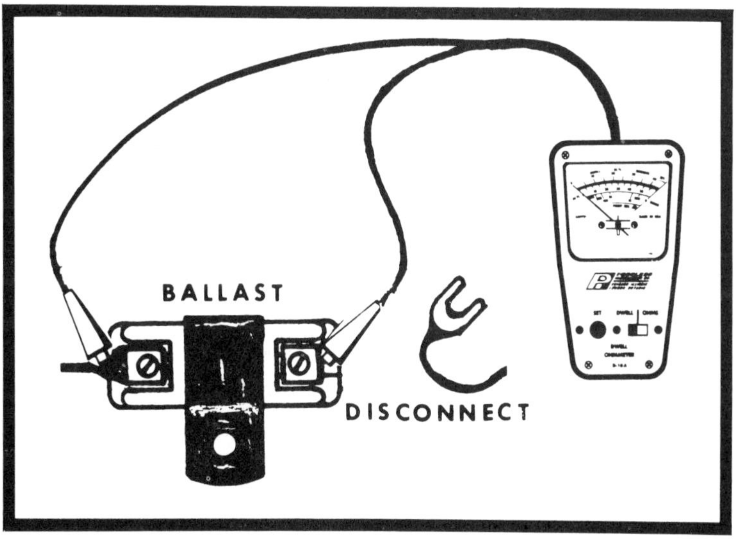

Fig. 1-13. Ohmmeter connected to an unknown resistance.

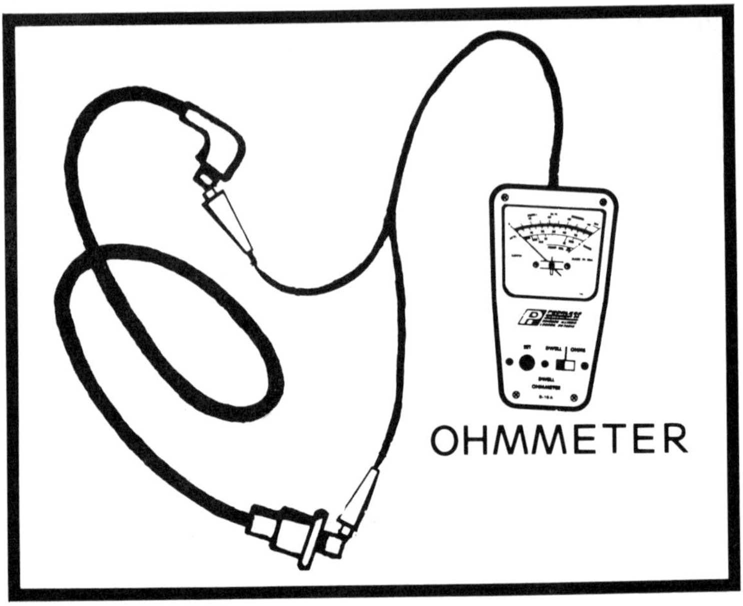

Fig. 1-14. Ohmmeter connected to spark plug wire.

either coil screw terminal. The other ohmmeter lead is clipped to a small screwdriver or prod and inserted into the coil tower. (See Fig. 1-16.)

The ohmmeter reading is compared with manufacturer's specification. The resistance of most coils will fall within the range shown in the table, Fig. 1-17.

TYPICAL RESISTANCE WIRE VALUES

WIRE TYPE	NOMINAL VALUE (Ohms Per Foot)	ACCEPTABLE VARIATION (Ohms Per Foot)
TVRS-LR		3000 - 7000
TVRS-HR		5000 - 12,000

Fig. 1-15. Table showing typical resistance values for spark plug wires.

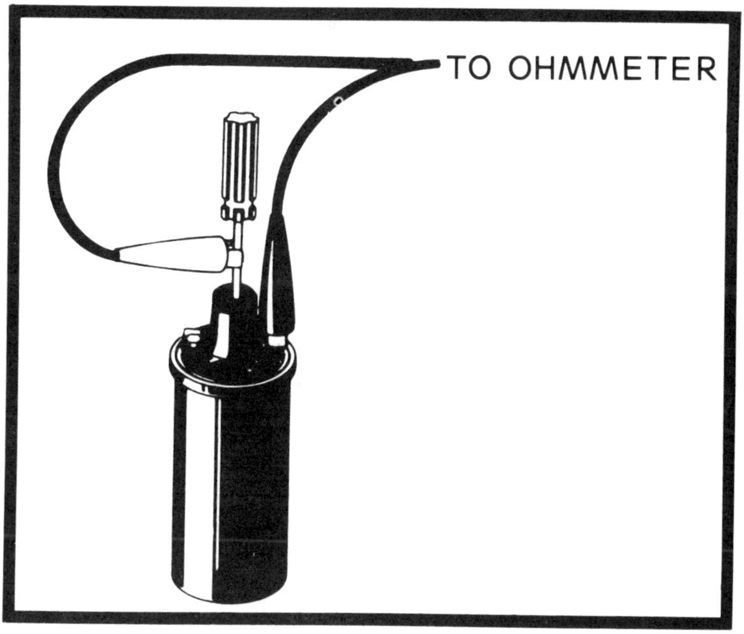

Fig. 1-16. Ohmmeter connected to measure coil secondary resistance.

Coil Ground

Disconnect both primary wires from the coil and connect one ohmmeter clip to a coil screw terminal and the other to the coil case as shown in Fig. 1-18. Set the ohmmeter to the high ohms position. The ohmmeter should show no movement. Any meter indication is evidence of a grounded coil winding.

MANUFACTURER	SECONDARY RESISTANCE (Ohms)
American Motors	6.5 K - 9.5K
Chrysler	9.5 K - 11.5K
Ford	7.5 K - 9.0K
Delco Remy	5.5K - 9.5K

Fig. 1-17. Table showing typical resistance values for coil secondary resistance.

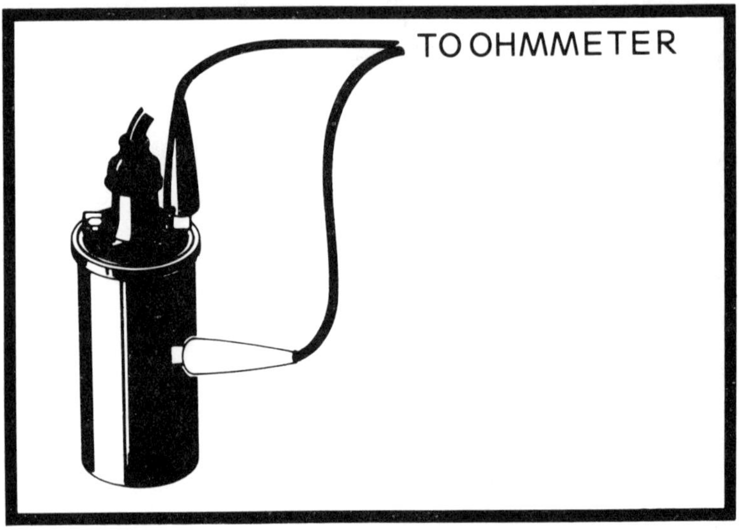

Fig. 1-18. Ohmmeter connected to measure resistance from coil primary to case.

Coil Primary Resistance Test

If any coil resistance tests are to be made while the coil is mounted in the vehicle, it must be electrically isolated from the vehicle's electrical system by disconnecting the primary leads from its terminals and removing the secondary lead from the coil tower.

The coil primary resistance test is performed with the ohmmeter in the low ohms position. The ohmmeter leads are connected to the primary side of the coil as shown in Fig. 1-19. The ohmmeter reading is compared with the specifications. No deflection of the ohmmeter indicates an open primary coil winding.

Condenser Test

The ohmmeter can be used as a quick test for condensers. This test will only indicate whether the condenser is open or shorted. The ohmmeter cannot be used for capacity measurements, nor can it be used to measure the series resistance or the actual leakage resistance of a condenser. The test is performed in the high ohms position. One side of the condenser is disconnected from the circuit, and the ohmmeter leads are connected as shown in Fig. 1-20.

Fig. 1-19. Ohmmeter connected to measure coil primary resistance.

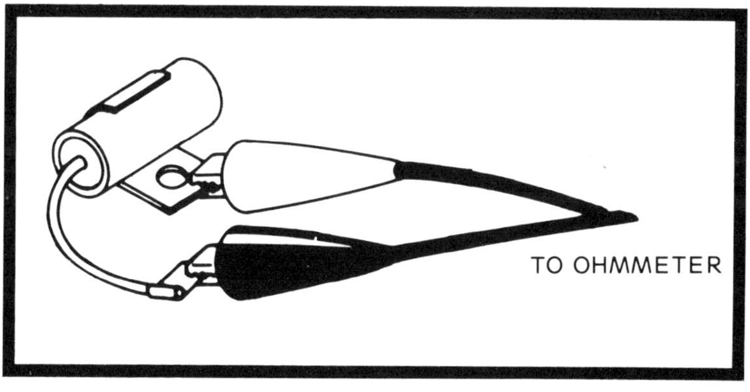

Fig. 1-20. Ohmmeter connected to test condenser.

If the condenser is shorted, the pointer will move to the right side of the scale (low resistance side) and stay there. If the condenser is good, the pointer will momentarily deflect toward the low resistance side and then return to the high resistance area. If a good condenser is to be tested again, momentarily touch the two condenser leads together before connecting the ohmmeter. This removes the charge placed on the condenser by the ohmmeter's internal battery so that it may be tested again.

Chapter 2

Battery-starter Tester

Since the instrument manufacturers frequently combine the testing procedures for the battery and starting circuits into one instrument, these two circuits will be considered together. The instrument for testing these circuits is called the battery-starter tester. After a brief description of the battery and battery maintenance, the battery-starter tester will be discussed in detail and the various tests it can perform will be illustrated.

BATTERY

The primary function of the battery is to provide power to operate the starting motor. It must also supply energy to the ignition circuit during the starting period, and must accomplish this even under the adverse condition of low temperature. In addition, the battery must also provide for a limited time, current to satisfy the electrical demands of the vehicle which are in excess of the output of the charging system.

Basically stated, a battery is composed of two dissimilar metals in the presence of a liquid called the electrolyte. The common auto battery cell is constructed of a series of positive plates of lead peroxide and negative plates of spongy lead. These plates are insulated from each other by means of separators. All the positive plates are interconnected and all the negative plates are interconnected. These interconnected series of positive and negative plates are submerged in the electrolyte solution composed of sulphuric acid and water.

The no-load voltage for each cell in the conventional automotive battery is 2.10 volts. Therefore, a 6-volt battery consists of three 2.10-volt cells, and a 12-volt battery consists of six 2.10-volt cells.

A chemical reaction takes place in the battery whenever the battery terminals are connected to a closed circuit. For example, during discharge the lead and lead peroxide plates are gradually changed to lead sulfate, and the sulfuric acid is

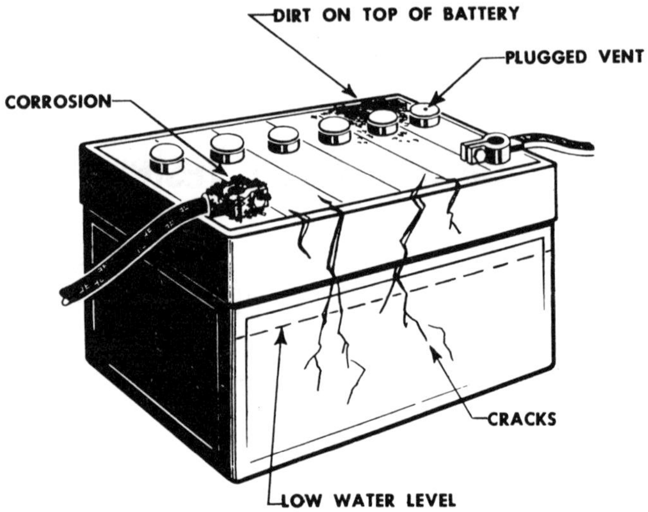

Fig. 2-1. Visual inspection of battery.

changed to water. At the same time, current flows out of the battery.

The chemical process is reversible; and whenever the battery is charged by the charging system in the car or by a battery charger, it returns to its initial state. That is, the positive and negative plates return to their original condition and sulphuric acid is reformed.

Batteries frequently are rated in ampere-hours. This rating is determined in a laboratory with the battery fully charged and at a temperature of 80 degrees F. The battery is discharged for twenty hours at a constant rate. At the end of the 20-hour period, the battery must have an average cell voltage of 1.75 volts or above. The battery is rated by the amperage it can supply under these conditions. For example, a battery supplying 3 amps would be given a 60 ampere-hour rating (3 times 20). A battery which can supply 6 amps would be rated at 120 ampere-hours.

Battery care and testing are relatively simple. Many undesirable conditions can be seen and corrected before they result in battery trouble. Therefore, the first test of a battery is visual inspection. The following items should be checked, see Fig. 2-1:

1. Inspect the battery case for cracks and leaks.

2. Inspect battery posts, clamps, and cables for breakage, loose connections, corrosion, and other faults.

3. Note whether or not the top of the battery is clean and dry. Dirt and electrolyte on top of the battery cause excessive self-discharge.

4. Be sure that the cell vents are open.

5. Be sure that the battery carrier is solidly mounted and in good condition and that the battery hold-down is properly tightened. A loose battery carrier or battery hold-down will allow the battery to be damaged by vibration and jarring. An excessively tightened battery hold-down may buckle or crack the battery case.

6. Inspect battery for raised cell covers or warped case which may indicate that the battery has been overheated or overcharged at some time.

7. Inspect electrolyte level. If electrolyte is below the top of the plates, add water.

After the visual inspection has been performed, the battery can be further tested with the aid of a hydrometer.

Hydrometer

The specific gravity of the electrolyte in a battery indicates the state of charge of the battery. A hydrometer is an

HYDROMETER TESTS
SPECIFIC GRAVITY READINGS

Specific Gravity	State of Charge
1.260-1.280	100 %
1.230-1.250	75 %
1.200-1.220	50%
1.170-1.190	25%
1.140-1.160	Very little useful capacity
1.110-1.130	Discharged

Fig. 2-2. Specific gravity reading obtained with hydrometer.

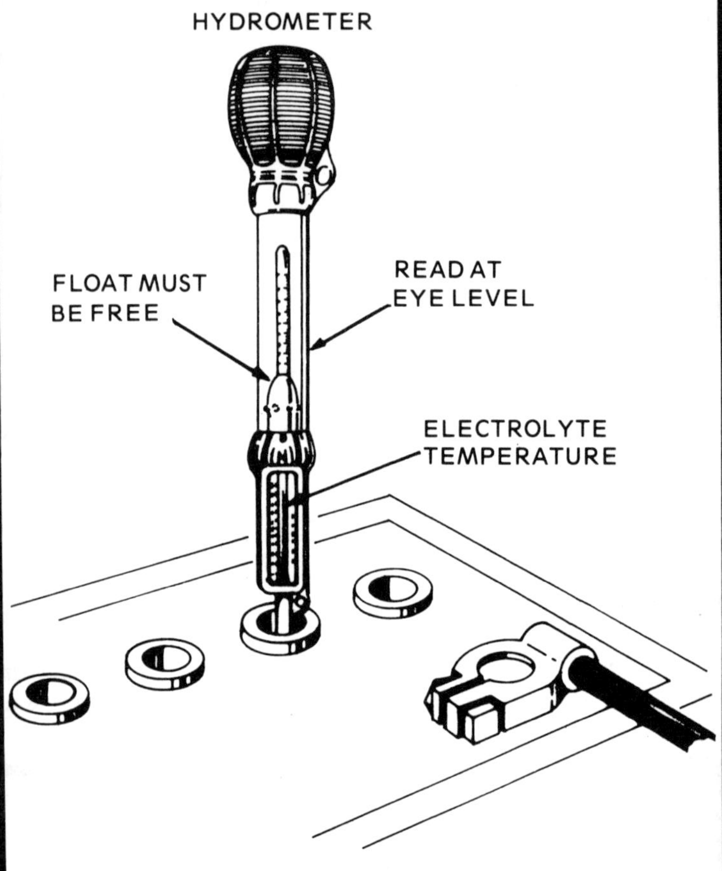

Fig. 2-3. Use of hydrometer.

instrument used to measure the specific gravity. If the state of charge of the battery is low, the hydrometer will read low. If the state of charge is high, the hydrometer will read high.

As an example, a reading from 1.260 to 1.280 indicates a fully-charged battery; a reading from 1.200 to 1.220 indicates a battery that is in a half-charged condition; readings below 1.200 indicate that the battery is in a discharged condition and cannot give satisfactory service (Fig. 2-2). A battery should be recharged whenever the specific gravity falls below 1.215.

The definition of specific gravity is the weight of a liquid compared to the weight of an equal volume of water. The specific gravity of chemically pure water at 80 degrees F is "one." A measurement of the specific gravity at any other temperature requires compensation of the reading. This is done by adding 4 points (.004) for each 10 degrees above 80 degrees F and subtracting 4 points for each 10 degrees that the electrolyte temperature is below 80 degrees F.

The hydrometer is used by removing enough electrolyte from one cell to allow the float to float freely without touching top or bottom (Fig. 2-3). Then by holding the hydrometer vertically to prevent the float from touching the sides, one should read the scale at eye level. Note that if water has just recently been added to the battery, the hydrometer readings will be erroneous.

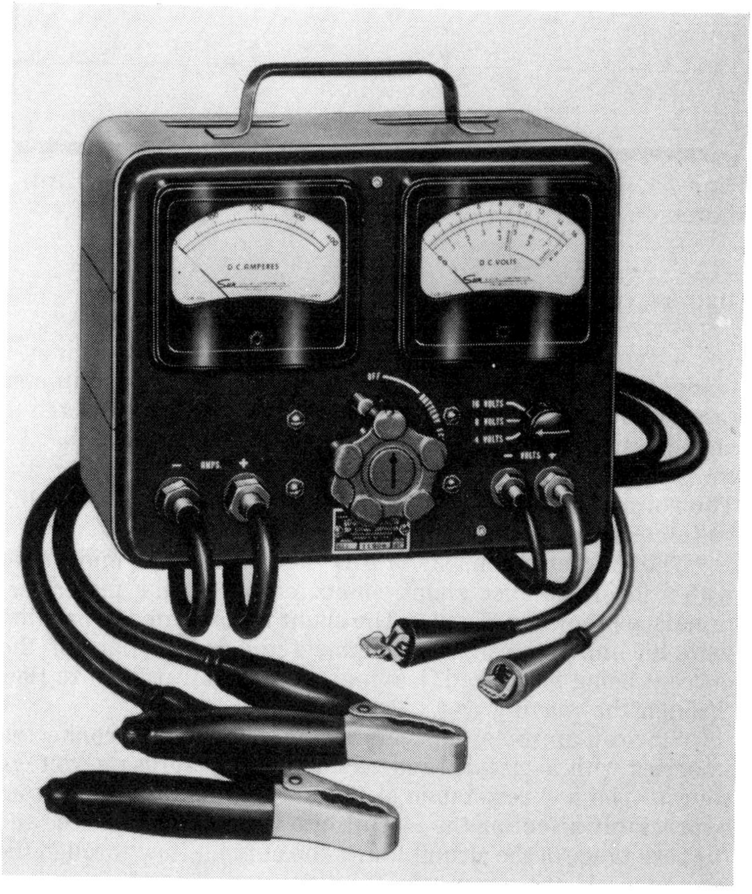

Fig. 2-4. Battery-starter tester.

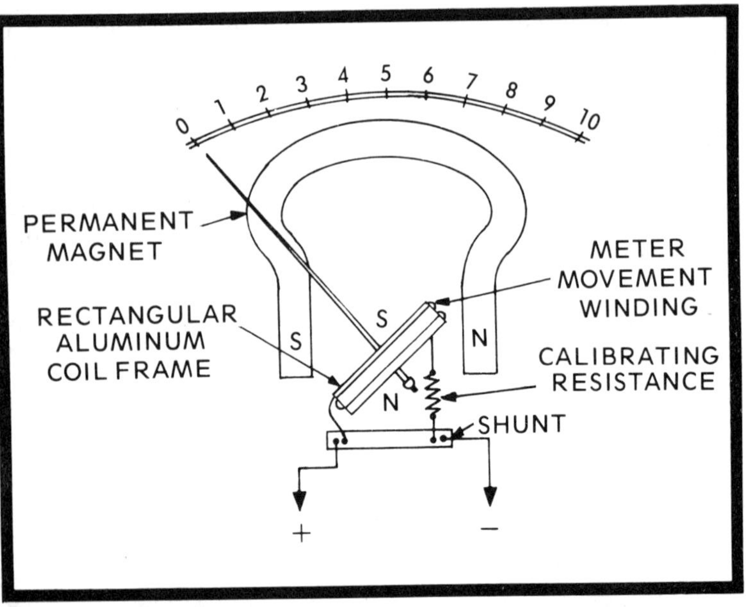

Fig. 2-5. D'Arsonval meter movement used as an ammeter.

Battery-Starter Tester

A typical battery-starter tester combines a voltmeter (ranges of 0-4 and 0-16 volts, or similar) with an ammeter range of 0-300 or 0-500 amperes. The ammeter is connected to an adjustable resistor, commonly called a carbon pile. An example of a typical battery-starter tester is shown in Fig. 2-4. The voltmeter in the battery-starter tester is identical to the voltmeter considered in the first chapter.

The ammeter consists of a D'Arsonval meter movement with a low-resistance shunt connected across the meter terminals, as shown in Fig. 2-5. The shunt is connected in parallel with the moving coil and conducts the major part of the current being measured, leaving only a small portion to flow through the moving coil.

Since an ammeter has very low resistance, it is connected in series with a circuit, and never directly across a voltage source. The low resistance of the ammeter prevents it from appreciably affecting the circuit into which it is inserted, and the resistance of the circuit limits the current flow through the ammeter. If the ammeter is connected directly across a voltage source, such as a car battery, it will be instantly damaged, Additional details on the ammeter will be found in the next chapter.

The adjustable resistor in the battery-starter tester is used to draw current from the battery. It is connected across the battery terminals. The ammeter in the battery-starter tester is internally connected in series with the adjustable resistor so as to measure the current drawn from the battery. The two leads from the tester have the adjustable resistor and ammeter in series but are simply called ammeter leads. The details on the construction of the adjustable resistor are considered in the next chapter. Several basic battery tests using the battery-starter tester will be explained later.

Battery Leakage Test

On a battery with a dirty top use a voltmeter set to its lowest range to test for leakage current that may self-discharge the battery. The leakage test is performed by clipping the negative voltmeter lead to the battery negative terminal. The positive voltmeter lead should be moved over the insulated surface of the top of the battery, see Fig. 2-6. Any reading is an indication of electrical leakage. This undesirable electrical path is composed largely of electrolyte which has been expelled from the battery through the fill cap vents by the action of the charging system.

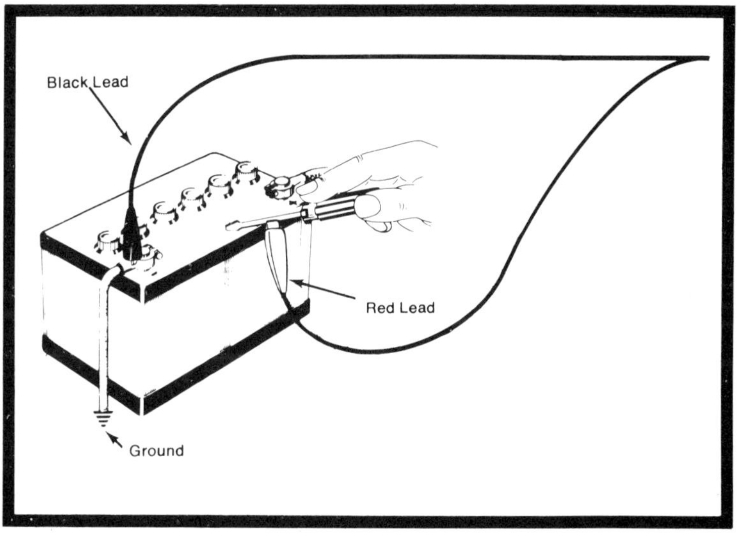

Fig. 2-6. Move voltmeter clip across surface of battery for battery leakage test.

If meter readings indicate any electrical leakage, the battery top should be washed with a solution composed of one spoonful of baking soda mixed in a pint of water. After the bubbling action induced by acid neutralization stops, the battery top should be rinsed with clean water.

No-Load Voltage Test

When a 16-volt voltmeter is connected across the terminals of a battery with all switches off (lights, ignition, etc.), the meter indicates the no-load battery voltage. The no-load voltage is a rough indication of the state of charge of the battery, but it should be noted that a hydrometer gives more accurate information on this condition. The no-load voltage of a fully charged 6-volt battery should be 6.3 volts or more. A fully charged 12-volt battery should read 12.6 volts or more.

It is not enough, however, for a battery to have a high no-load voltage and to be in a charged condition. The battery must also be able to maintain a certain minimum voltage while it is delivering a heavy load to the starter motor. If the battery is not able to maintain this minimum voltage while cranking the starter motor, it does not have sufficient capacity to guarantee ignition. It therefore becomes necessary to perform the battery capacity test to determine whether the battery can provide sufficient current while maintaining a specified minimum battery voltage.

Battery Capacity Test

The capacity of a battery is the ability of the battery to furnish current and maintain a minimum necessary voltage when placed under a load. A battery that passes the capacity test will provide dependable performance.

The battery capacity test should not be performed on newly activated dry-charge batteries or on stock batteries that have been standing idle. Recently activated dry-charge batteries can be tested after they have been in service for several days and the acid has had an opportunity to fully penetrate the pores of all the dry-charged plates. Stock batteries may be tested if they have been freshly charged within the preceding 24 hours. Unused or standing batteries quickly accumulate a thin film of plate sulphation, which, although harmless and quickly removed in normal car operation, tends to give an inaccurate reading of the battery condition with any type of load test.

Two methods of performing the battery capacity test are in common use. The first method uses the car's starting system as a load. The second method makes use of the carbon pile load in the battery-starter tester. The procedures for both methods are described below:

A. Starting System as the Load

1. Before the test is performed, the battery must be fully charged. Each cell should have a minimum reading of 1.240 on the hydrometer.

2. A battery loses much of its capacity at low temperatures. Therefore, this test should not be performed at extremely low temperatures.

3. Connect a voltmeter across the battery terminals.

4. Ground the secondary coil wire so that the engine will not start.

5. Crank the engine for 15 seconds and read the voltage during the last second.

6. On 12-volt systems, a reading of 9.6 volts or above indicates a good battery; a reading below 9.6 volts indicates a defective battery. On 6-volt systems, a reading of 4.8 volts or above indicates a good battery.

B. Carbon Pile as the Load

1. Connect the voltmeter and ammeter leads from the battery-starter tester across the battery as shown in Fig. 2-7.

2. Turn the control knob on the carbon pile until the ammeter reading is equal to 3 times the ampere-hour rating of the battery. (Example: 180 amperes for a 60 AH battery.)

3. Maintain the load for 15 seconds and note the voltmeter reading. Then turn the carbon pile knob to the off position.

4. The voltage should not drop below 9.6 volts for a 12-volt battery, nor below 4.8 volts for a 6-volt battery. Voltage readings below these values indicate a defective battery.

If the voltage falls below the values given, either the battery is in a low state of charge or it is in poor condition and will soon fail. To find out which is the case, charge the battery at a fast rate (40 amperes for a 12-volt battery or 75 amperes for a 6-volt battery) for at least three minutes; then note the voltage reading across the battery terminals while the charger is still operating. If the voltage reading exceeds 15.5 volts on a 12-volt battery or 7.75 volts on a 6-volt battery, the battery is sulphated or worn out.

If the voltage reading is less than 7.75 volts on a 6-volt battery or 15.5 volts on a 12-volt battery, the battery is

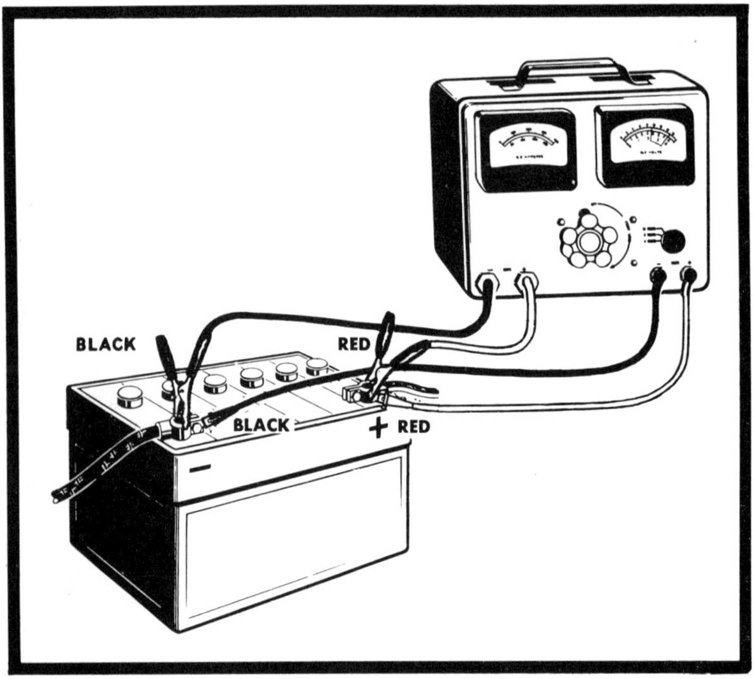

Fig. 2-7. Lead connections for performing the battery capacity test.

probably just discharged. In some rather rare instances the low voltage across the battery could be due to a defective cell.

Battery Charging

The battery charger shown in Fig. 2-8 will fast or slow charge and is polarity protected to prevent damage to both the battery and battery charger if connected wrong. A battery that is in good condition but requires recharging will accept a reasonably large amount of charging current without harm. This type of battery may be charged quickly at a high rate with a battery charger on fast charge.

A battery that has become sulfated, however, will not accept a high rate of charging current without possible damage. A sulphated battery is one which has hardened plates as a result of standing in a discharged or half charged condition for long periods. This condition can be caused by undercharging due to a defective charging system or high resistance in the charging circuit. A battery in a sulphated condition provides increased resistance to current flow within

Fig. 2-8. Fast and slow charge battery charger.

the battery. Flow of a high rate of charging current through this resistance creates heat, which can result in boiling of the electrolyte, warping of the plates, and eventual damage to the separators.

A sulfated battery should be charged over a long period at a low rate. In this manner, sulfate formation on the plate may be gradually broken down and the battery returned to its normal state. Badly sulfated batteries may take from 60 to 100 hours to recharge completely.

A fast charge rate is approximately 40 amperes for a 12-volt battery and approximately 75 amperes for a 6-volt battery. The normal fast charge rate is from ¼ to ¾ hours. A slow charge rate is 7 percent of the battery ampere-hour rating or less. The normal slow charge period is from 12 to 24 hours.

The battery is only about 80 percent charged after a fast charge. To charge the battery further requires slow charging. A battery is considered fully charged when three consecutive hydrometer readings, taken at hourly intervals while the battery is being charged, show no rise in specific gravity.

Be sure to observe all precautions relative to working around a battery while it is being charged. Explosive hydrogen gas is liberated from the electrolyte while the battery is being charged. Sparks from a lighted cigarette or from charger clamps being disconnected while current is still flowing may cause an explosion that will destroy the battery and possibly inflict personal injury.

Remember never to overcharge. Overcharging causes a loss of water in the cells and also causes the battery to heat up inside, which oxidates the positive plates and results in a loss of cell capacity and early failure.

STARTING CIRCUIT

The starting system is used to crank the engine until it can continue running under its own power. The starting system (Fig. 2-9) consists of an electric motor to crank the engine, a battery to furnish the electrical energy to operate the cranking motor, and a switch to complete the circuit between the battery and the motor. This switch usually is an electrically operated device called a solenoid.

The starter is a high-speed electric motor, which, through large gear reduction, provides sufficient torque to crank the engine at 140 to 250 rpm for starting. The starter is engaged with the engine flywheel by means of either a starter solenoid or a Bendix type starter drive.

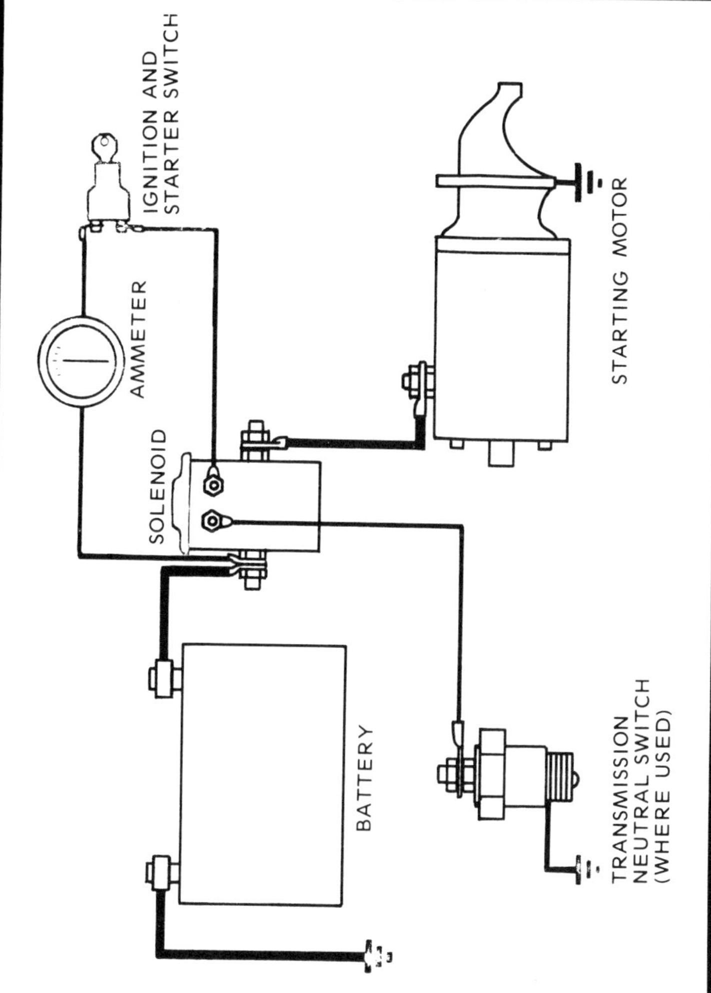

Fig. 2-9. Starting circuit.

A starter solenoid is an electromechanical device having a heavy coil around a plunger that is connected to a shift lever. Closing the starter circuit activates the solenoid as an electromagnet. This action pulls the plunger and moves the shift lever, through suitable linkage, to engage the starter pinion with the gear teeth on the engine flywheel. An over-running clutch disengages the starter pinion as soon as the engine starts.

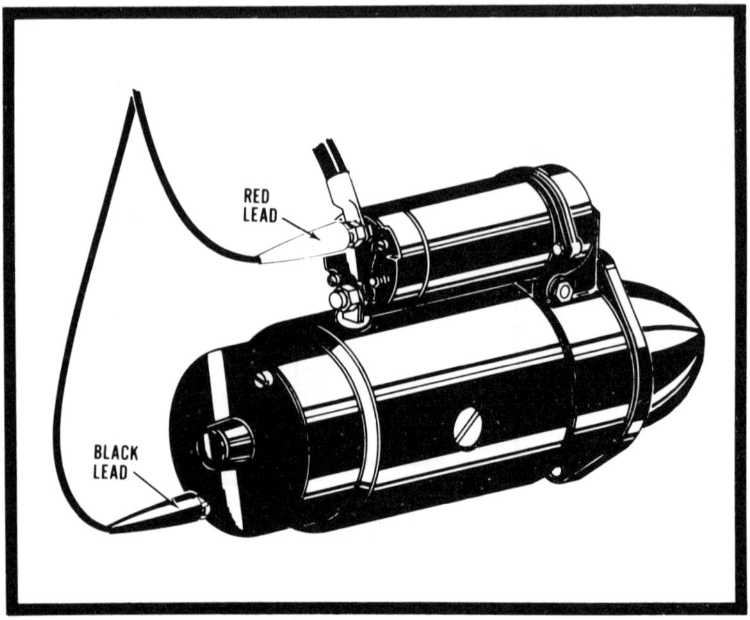

Fig. 2-10. Lead connections for starter motor voltage test.

The Bendix type starter drive uses inertia to engage the starter motor with the engine flywheel. The starter spins a shaft at high speed, which causes a pinion to travel down the shaft until it meshes with the gear teeth on the engine flywheel. As soon as the engine starts, the flywheel travels faster than the pinion is being turned by the starting motor. This causes the pinion to turn in the opposite direction on its threaded shaft, which automatically disengaged it from the flywheel.

Starter Motor Voltage Test

A fully charged battery must be in the vehicle before this or any other starter circuit test can be made. In addition, the starter motor should first be inspected for obvious defects such as loose connections or a loose mounting.

With the voltmeter in the 16-volt position, connect the positive lead to the starter terminal or the starter switch terminal. Connect the minus voltmeter lead to the starter motor housing (ground), see Fig. 2-10. Remove the coil wire from the distributor cap and ground it so that the engine will not start.

Crank the engine while reading the voltmeter. The voltage should be 4.7 volts or higher for 6-volt batteries, and 9.4 volts or higher for 12-volt batteries. If the voltage at the starter is sufficiently high and if the starter motor cranks the engine with good speed, the starting motor circuit is in good order, and no further tests are necessary. But, if the cranking speed is poor or if the voltage at the starter is low, check for poor conductivity in the parts which make up the circuit by performing the starter circuit ground cable and power cable tests.

Starter Circuit Ground Cable Test

Set the voltmeter to the low volts scale. Connect the voltmeter leads as shown in Fig. 2-11. Notice that the minus test lead is connected directly to the battery post. It must not touch the cable connector. Use a screwdriver or awl in the manner shown in the illustration to avoid touching the connector and to make a good connection directly to the battery post. The positive test lead is connected directly to the starting motor housing.

With the secondary coil wire from the distributor removed and grounded, crank the starter and take a reading on the voltmeter. The voltage loss as read on the meter should be 0.2 volts or less.

If the reading exceeds the allowable amount, the exact location of the loss must be found. High voltage losses can be

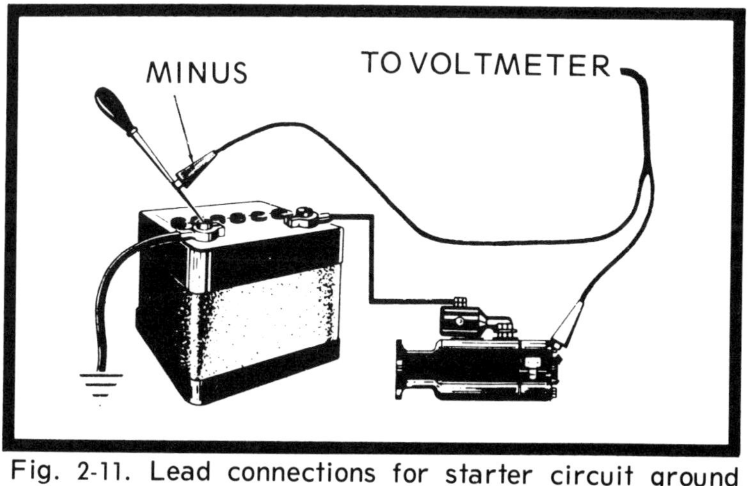

Fig. 2-11. Lead connections for starter circuit ground cable tests.

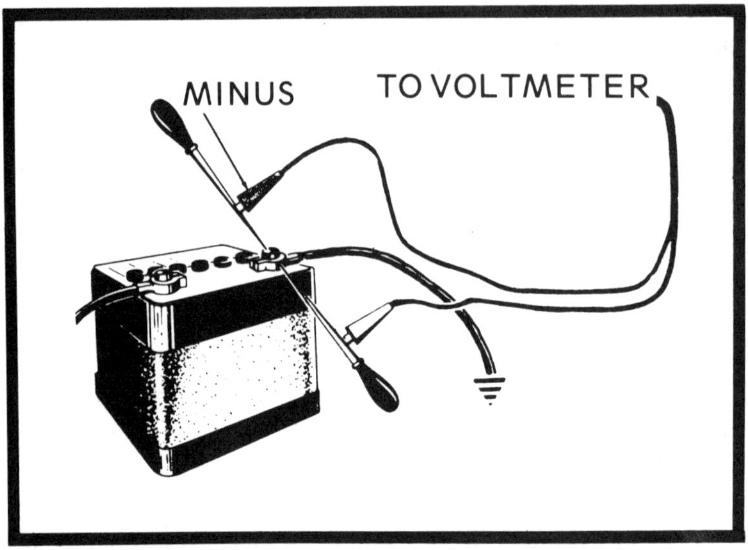

Fig. 2-12. Testing connection of ground strap to battery post.

caused by dirty or loose connections, defective cables, or cables too small to carry the current.

For example, the problem could be caused by a poor connection between the battery post and the ground strap. To determine the condition of this junction, connect one test lead to the battery ground post and the other to the terminal as shown in Fig. 2-12. Close the starter switch. A reading over 0.1 volt indicates that this junction is not as clean and tight as it might be.

If the junction at the battery post is good, test the junction on the other side of the ground strap where it is grounded to the engine block. To do this, place one test lead on the cable connector and the other on ground. Close the starter switch. A reading over 0.1 volts indicates too much loss through the junction under test.

If both junctions are good, the cable itself might be at fault. If a ground strap or cable has broken strands, it should be replaced even without testing. Frequently a perfect looking cable is defective; the cable may be undersize or the connectors on either end may not be soldered or fastened properly to the cable strands.

To test the cable for voltage drop or power loss, connect one test lead to each end and close the starter switch. A reading over 0.1 volts indicates a defective cable.

Starter Circuit Power Cable Test

Connect the plus voltmeter lead to the positive battery post and the minus lead to the starter terminal on the starter, see Fig. 2-13. With the secondary coil wire removed and grounded, crank the engine and read the voltage drop on the low volts scale.

The voltage reading should be 0.6 volts or less. If the reading exceeds this allowable value, measure each connection and cable until the trouble is located. The loss across each cable or across a switch should not exceed 0.2 volts.

Starter Current Draw Test

This test is performed with an ammeter connected in series with the battery and read while the engine is cranking. The secondary coil wire is grounded so that the engine will not start.

The typical automobile starter will draw between 120 and 240 amps, in warm weather, depending on the size of the engine. Unusually high current draw (260 amps or more), often accompanied by slow cranking speed, indicates a shorted circuit or that the armature is rubbing against the pole shoes of the starter motor. Unusually low current draw, often accompanied by slow cranking speed, indicates high

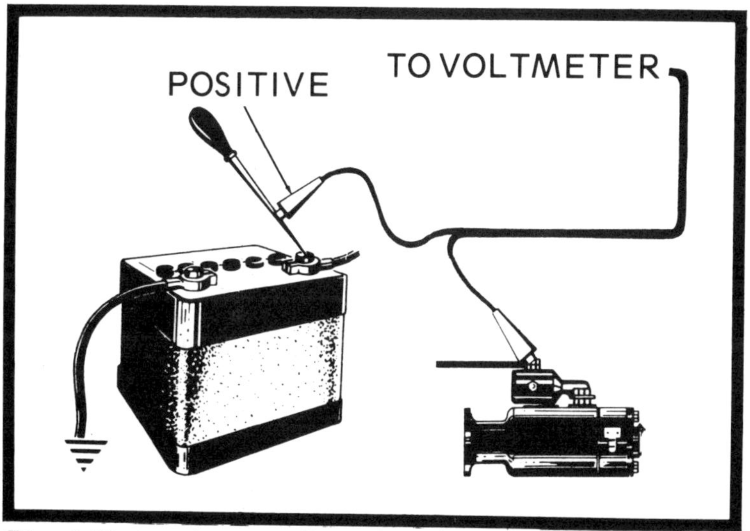

Fig. 2-13. Lead connections for starter circuit insulated cable test.

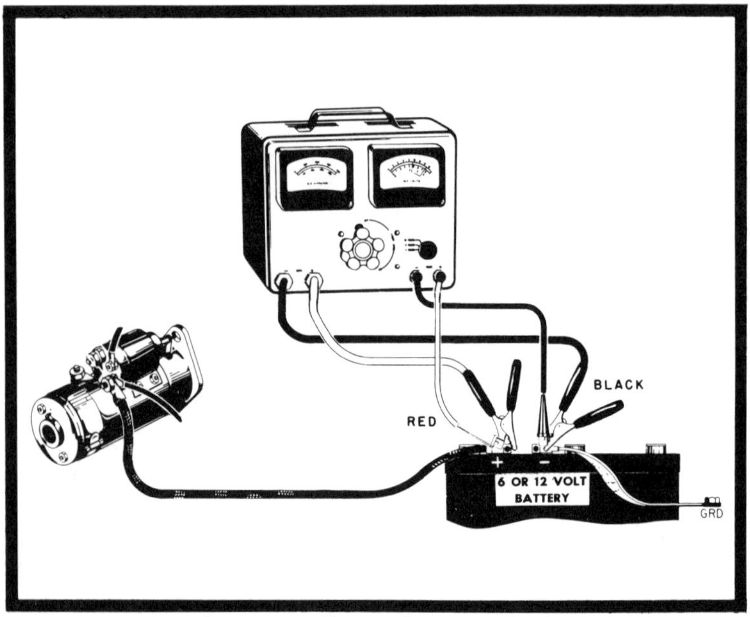

Fig. 2-14. Lead connections for using battery-starter tester for performing starter current draw test.

resistance within the starter motor (provided external circuits are known to be good) possibly caused by poor brush or commutator condition or by a poor connection in the field or armature circuits.

An alternate method of measuring starter current draw utilizes the battery-starter tester. The voltmeter and ammeter leads are connected across the battery terminals as illustrated in Fig. 2-14. The knob on the carbon pile which is in series with the ammeter is turned to off.

With the ignition disabled so that the engine will not start, crank the engine just long enough to note the exact reading on the voltmeter. Then by means of the knob on the carbon pile, insert enough resistance so that the same voltage is read as when the engine was cranking. The ammeter will now indicate the same number of amperes that the cranking motor required.

When a very quick check on starter current is required, the starting current indicator shown in Fig. 2-15 can be used. This instrument indicates by being held close to a cable which is carrying a high current. Heavy currents of 150 to as high as 400 amperes create a fairly powerful magnetic field, the

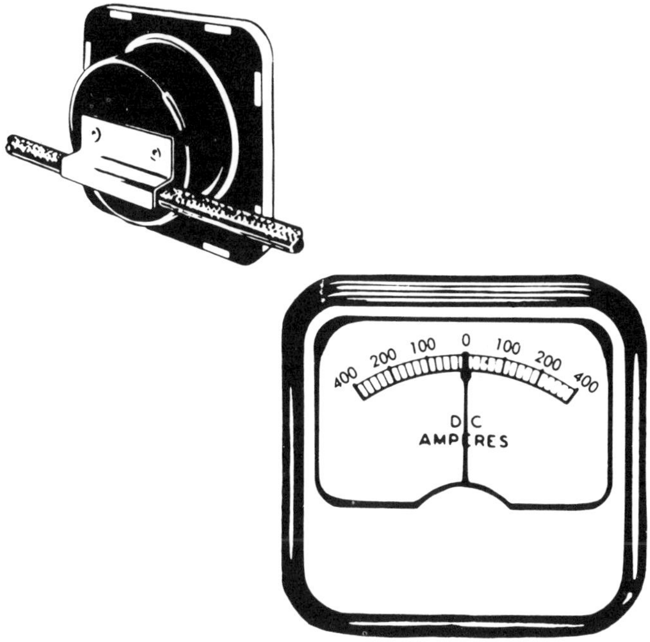

Fig. 2-15. Starting current indicator.

strength of which is proportional to the current. This field actuates the indicating device in the instrument and moves the indicating needle accordingly. Although this method of obtaining starter current draw is not as accurate as the others, it is sufficient for determining the condition of the starter.

Chapter 3

Charging System Analyzers

The purpose of the charging system is to maintain the storage battery in a charged condition or to recharge it when it has become discharged. In addition, the charging system supplies current for the lights, ignition, radio, heater, and other electrical accessories when the engine is operating. The charging system consists of four important components; battery, alternator or generator, regulator, and leads, see Figs. 3-1 and 3-2.

The battery and battery testers were discussed in the previous chapter. Here the emphasis will lie on the remaining components of the charging system and the methods used to test them.

ALTERNATORS AND GENERATORS

Either a generator or an alternator converts mechanical energy supplied by the engine into electrical energy. The alternator, however, possesses the ability to produce a current output at engine idle speed. This factor makes it a superior charging unit to the generator. A generator must be rotated at a considerable speed before a current output is developed.

The output of an alternator system is alternating current, which means that half the total output current flows in one direction and the other half flows in the opposite direction. Since alternating current cannot be used directly by the automotive electrical system, it must be converted to direct current. This is accomplished by passing the alternating current through diode rectifiers, which allows current to flow in only one direction.

The output of a generator system is direct current and, therefore, the diode rectifiers are unnecessary.

A and B Circuits

Charging systems are either A or B type. These differ only in the way in which the field circuit is connected and grounded.

In the A circuit (externally grounded field) the field circuit is connected to the armature terminal and grounded outside the generator or alternator through the regulator contacts. In the B circuit (internally grounded field) the field circuit is connected to the armature circuit through the regulator and grounded inside the generator or alternator.

GENERATOR REGULATOR

The generator regulator is the control unit for the generator. Its function is to automatically control generator voltage, and current output to safely meet all conditions of speed and electrical load. The generator regulator assembly usually consists of three units:

1. Cutout or reverse current relay
2. Voltage regulator
3. Current regulator

Cutout Relay

The cutout relay or circuitbreaker is used to open and close the circuit between the generator and battery. There is no output from the generator when the engine is stopped; consequently, if the battery-to-generator circuit remained closed, current would flow from the battery through the generator to ground, thereby discharging the battery. The cutout relay prevents such current flow by opening the generator circuit when the engine is stopped or is operated at a speed at which the generator output voltage is below battery voltage. Only when the generator output voltage exceeds that

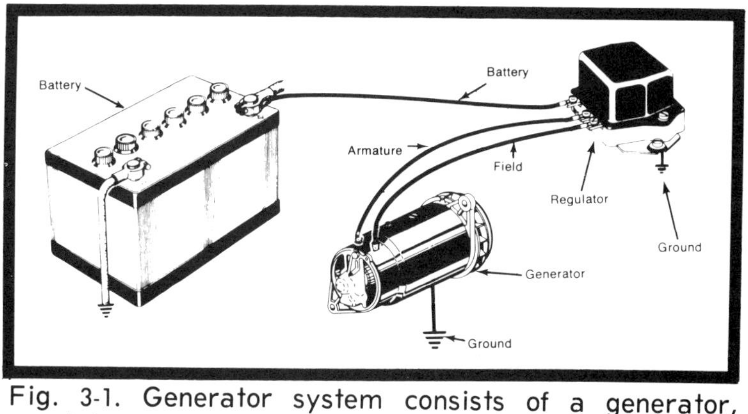

Fig. 3-1. Generator system consists of a generator, regulator, and battery.

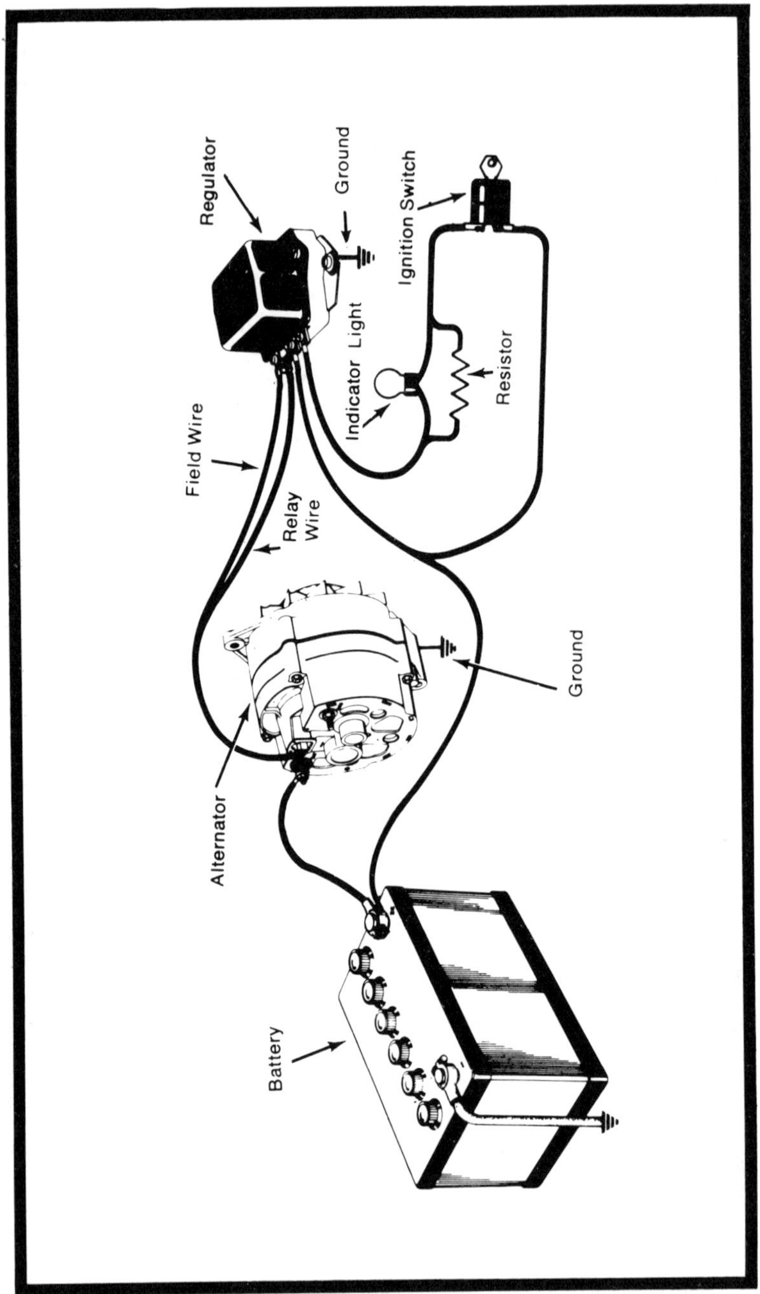

Fig. 3-2. Alternator system consists of an alternator, regulator, and battery.

of the battery will the circuit close and permit the generator to charge the battery. The two specifications important in checking a cutout relay are the closing voltage and opening current.

Generator Voltage Regulator

The voltage regulator is used to regulate the generator voltage to a predetermined value, thus protecting the battery and all vehicle electrical accessories from the damage of high voltage. When the generator voltage rises to a value at which the voltage regulator is set, the regulator contacts separate forcing the generator field current to flow through a resistor to ground. The resistor lowers the current flow in the generator field coils which automatically reduces the generator output.

Current Regulator

The current regulator is used to limit the generator current output to a predetermined value, thus protecting the generator against overload. When the generator output reaches its predetermined safe maximum output, the regulator contacts separate and insert a resistance in the field circuit which reduces the voltage and consequently the amperage output of the generator.

ALTERNATOR REGULATOR

The alternator regulator ordinarily contains the following components:

1. Voltage regulator
2. Field relay
3. Indicator light relay

The cutout relay used in the generator system is not needed in the alternator system because the diode rectifiers in the alternator system permit the flow of current in one direction only. These diodes will not allow battery discharge by current to flow from the battery into the alternator.

The current regulator used in the generator system is also not needed in the alternator system since the alternator is self-limiting in current output as long as voltage control is maintained.

Alternator Voltage Regulator

Every alternator system is controlled by a voltage regulator. The battery initially supplies the current for the alternator field coil. At this time, and during idle and low-speed operation, there is no voltage control problem. However, as alternator speed increases with engine speed, the voltage increase developed by the alternator will be imposed on the field coil. This increases the field strength and further raises the voltage output. Unless this voltage rise condition is kept under control, the high voltage developed will result in damage or shortened life expectancy of light bulbs, relay coils, turn signal flashers, breaker points and other voltage-sensitive units. The means of control is similar to the generator voltage regulator.

Field Relay

The field relay is used to energize the alternator field coil by connecting the coil to the battery when the ignition switch is turned on. Then the ignition switch is turned off, the field relay opens the circuit between the field coil and the battery to prevent battery discharge.

Indicator Light Relay

The indicator light relay is used in the alternator system when a light is mounted on the instrument panel to replace the ammeter. The relay permits the light to go on when the alternator is not producing voltage and causes the light to go out when voltage is produced. Illumination of the light when the engine is running indicates charging system trouble.

CHARGING SYSTEM TESTING

All components in a charging system are important. In order for the charging system to function properly, the generator or alternator must be capable of producing an output at least equal to its electrical rating; the regulator must be able to limit the generator or alternator output at a predetermined setting; the leads should be in good condition and the connections clean and tight. With high resistance in the leads and connections, the system cannot perform properly.

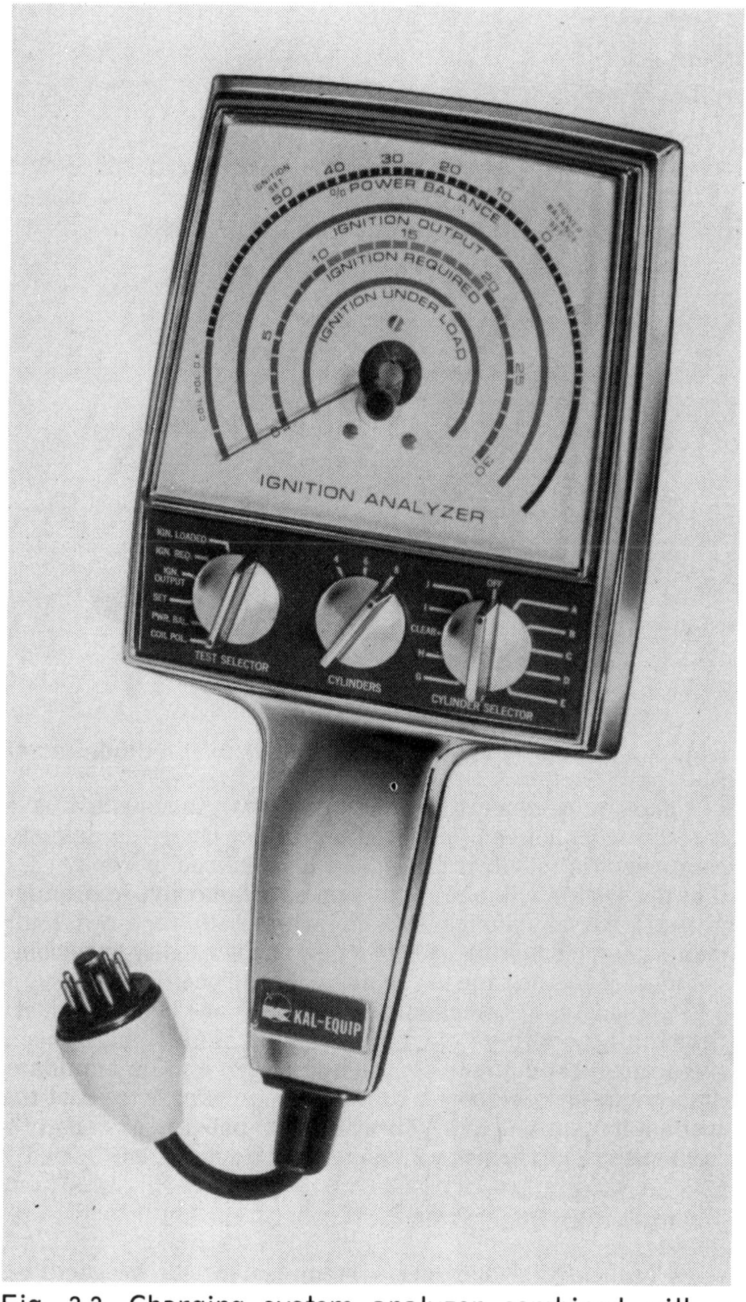

Fig. 3-3. Charging system analyzer combined with a starting system analyzer.

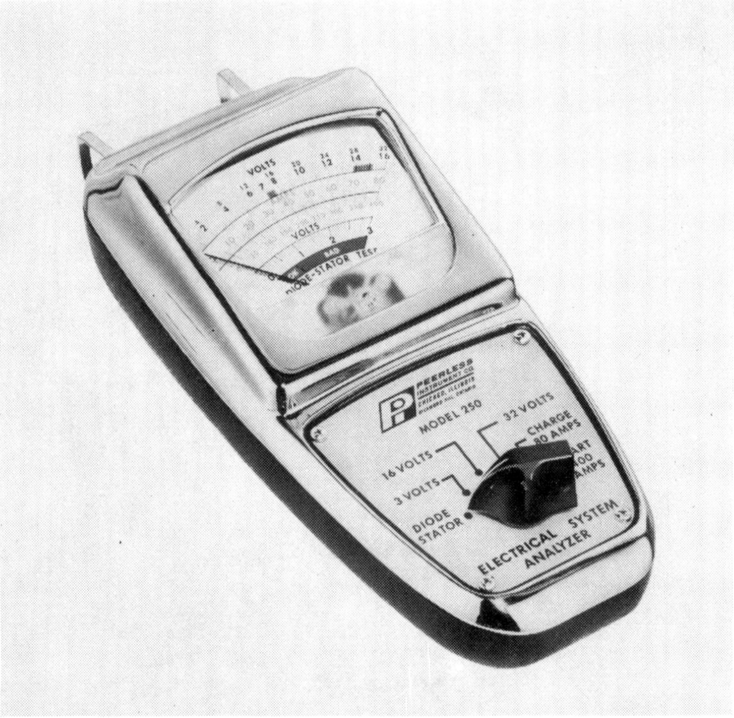

Fig. 3-4. Charging system tester with built in diode tester.

The main concern in testing a charging system is to determine which component is defective. Once the defective component is found, it can easily be repaired or replaced so that the system will again function satisfactorily. In practice, professional mechanics seldom attempt to repair a faulty component because the cost of a new one is usually low enough not to warrant the time in repairing the old one.

Obviously, the mechanic can replace one component at a time until the system starts functioning again; but, by using the proper test equipment, the guesswork can be eliminated and much time can be saved. More important is the fact that test equipment will tell you which components are really at fault so that the repairs you make will be permanent.

Charging System Testers

Frequently a charging system tester will be combined with a starting circuit analyzer as is shown in Fig. 3-3 since both circuits require a voltmeter and ammeter for testing and troubleshooting. Figs. 3-4 and 3-5 show other typical charging system analyzers.

Basically, a charging system analyzer consists of the following:

1. Voltmeter
2. Ammeter with battery post adapter

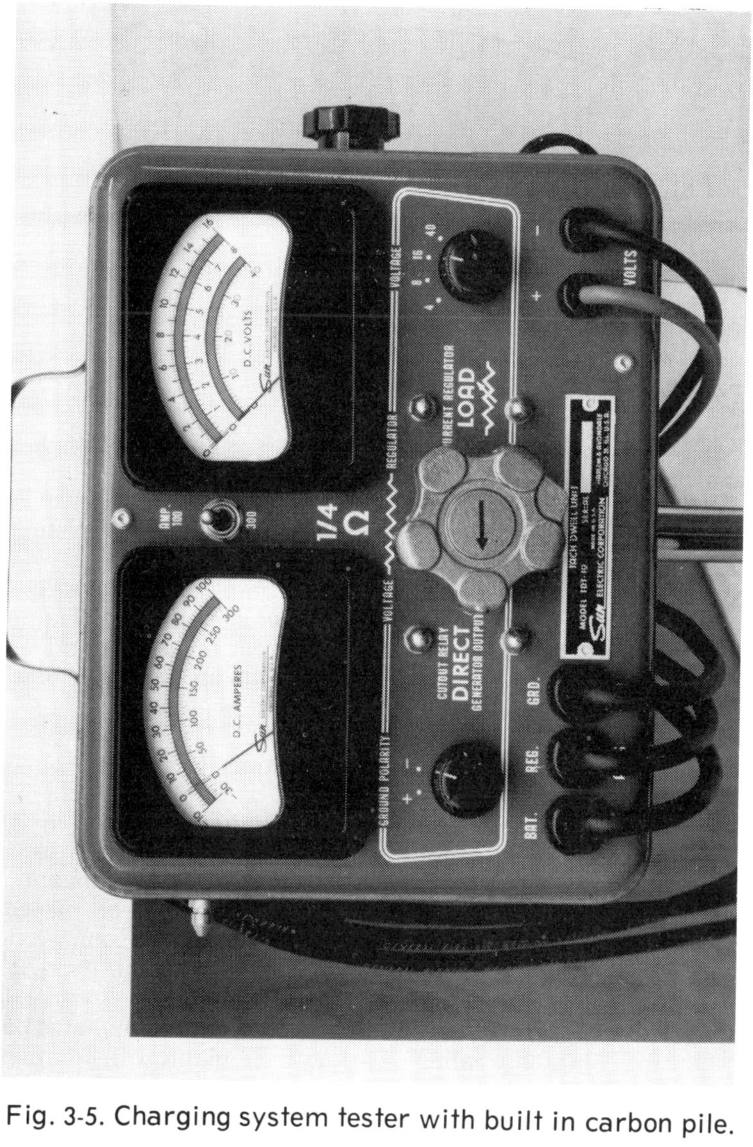

Fig. 3-5. Charging system tester with built in carbon pile.

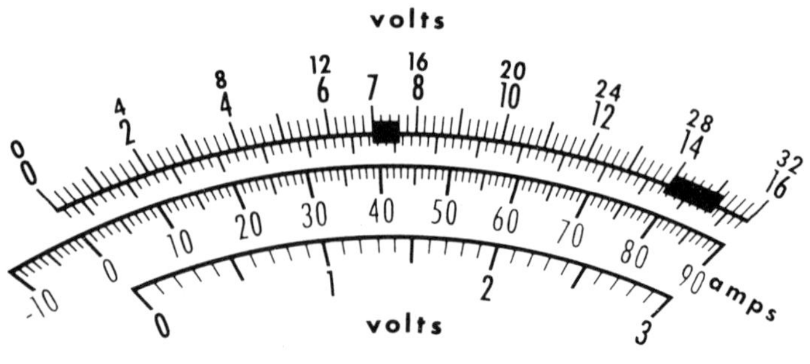

Fig. 3-6. Scale on typical charging system tester.

3. Resistance unit
4. Leads
5. Load unit or carbon pile

Voltmeter and Ammeter Scales

The scales of a typical charging system analyzer are shown in Fig. 3-6. The 16-volt scale is used for both 6- and 12-volt systems. The dark bars from 7 to 7.6 volts and from 14 to 15.2 volts are the areas in which most good 6- or 12-volt system voltage regulators are set. The 32-volt scale is used for 24-volt systems. The 3-volt scale is used for making voltage loss measurements regardless of the type of system. The 90-amp scale is used for making charging current measurements on all charging systems. Finally, the 0- to -10-amp scale is used for making cutout relay tests on generator systems.

Ammeter Shunt and Battery Post Adapter

In order for the ammeter to function properly, it is necessary to use an ammeter shunt. Fig. 3-7 illustrates a typical ammeter shunt which is usually made from manganin. As current flows through the shunt, a voltage is developed across the two points to which the meter leads are connected. The meter actually reads this developed voltage, but its scale is calibrated in amperes.

The ammeter shunt is either combined with or mounted on a housing commonly called a battery post adapter. In addition to the ammeter shunt, a resistance unit (¼ ohm) is also mounted on the battery post adapter. The resistance is used to

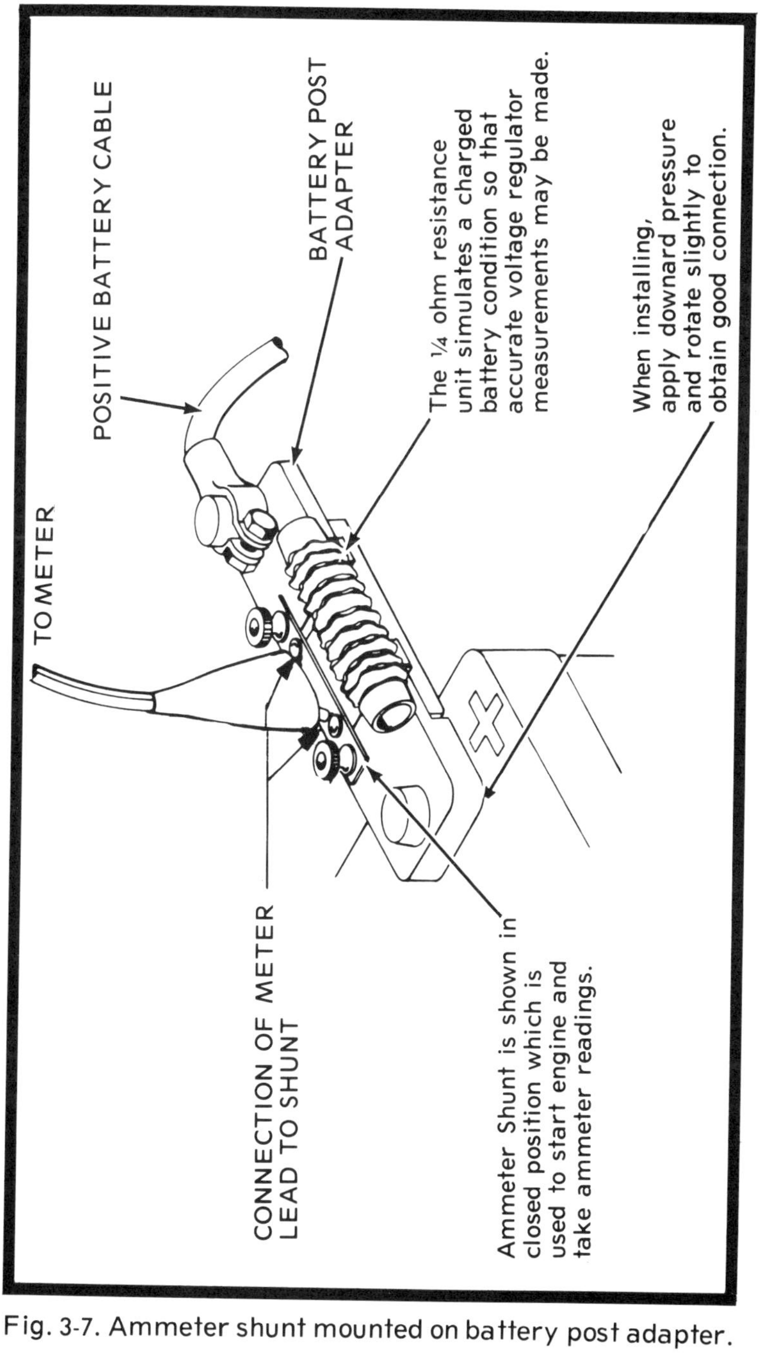

Fig. 3-7. Ammeter shunt mounted on battery post adapter.

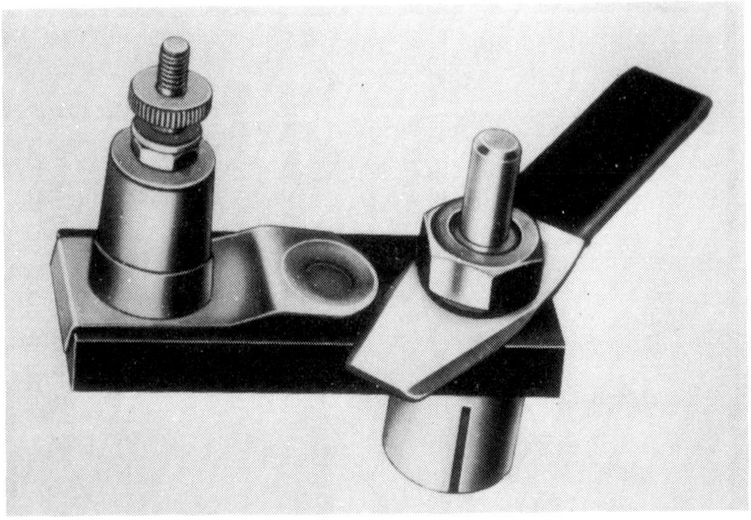

Fig. 3-8. Battery post adapter with switch.

simulate a charged battery so that accurate voltage regulator measurements may be made. By adding this resistance into the charging circuit, the amount of charging current is limited even when a completely discharged battery is placed in the circuit. By simulating a fully charged battery condition, the charging system voltage will rise to a value high enough to cause the voltage regulator to function, permitting its operating level to be measured.

The engine cannot be started with the ¼-ohm resistance unit in the circuit; therefore, it is necessary to short out the resistance unit with either the ammeter shunt as shown in Fig. 3-7 or with a special switch as shown in the battery post adapter design illustrated in Fig. 3-8. After the engine is running, the ammeter shunt can be removed or the switch can be opened, thereby inserting the ¼-ohm resistance into the circuit.

Load Unit and Carbon Pile

A load unit is illustrated in Fig. 3-9 and a carbon pile, which is usually built into the tester, is shown in Fig. 3-10. Both the load unit and the carbon pile are used to force the charging system to deliver full output current. These units are similar in that they are low resistance loads. The load unit is a fixed resistance made from a resistance wire. A switch on the load

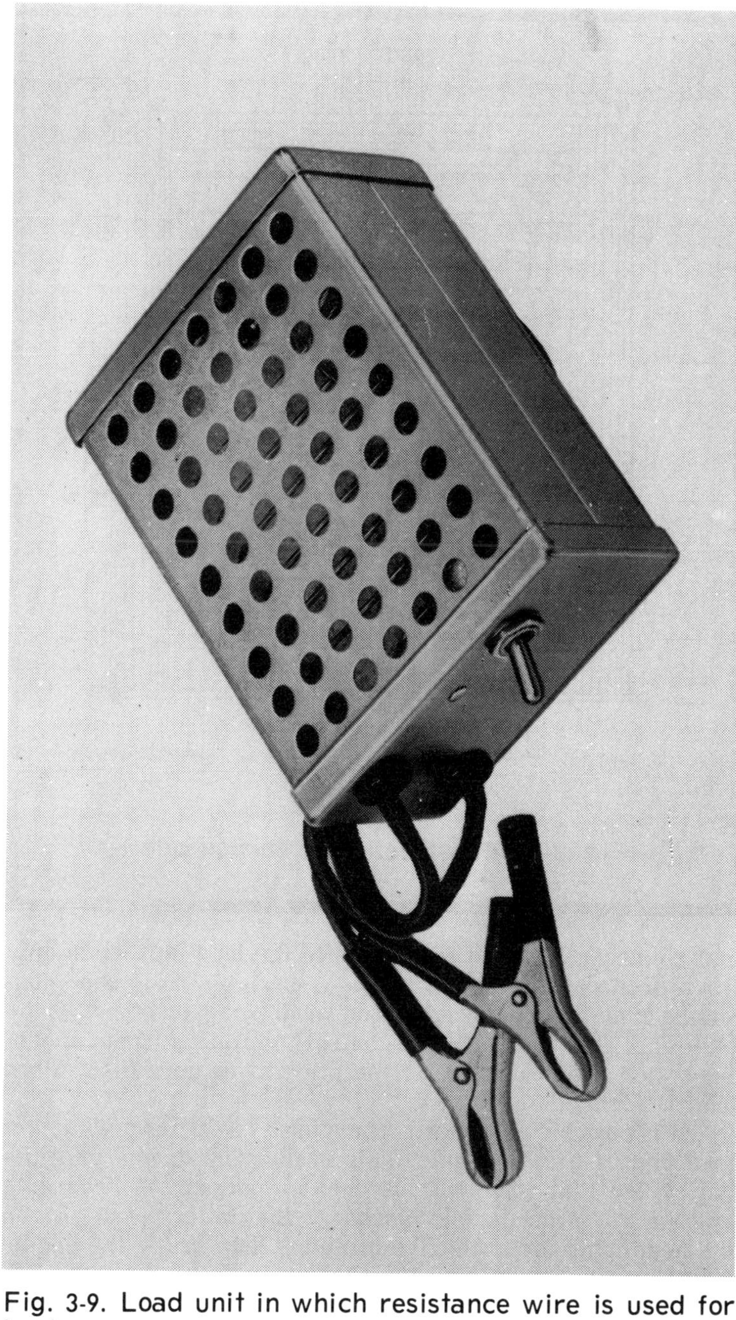

Fig. 3-9. Load unit in which resistance wire is used for load.

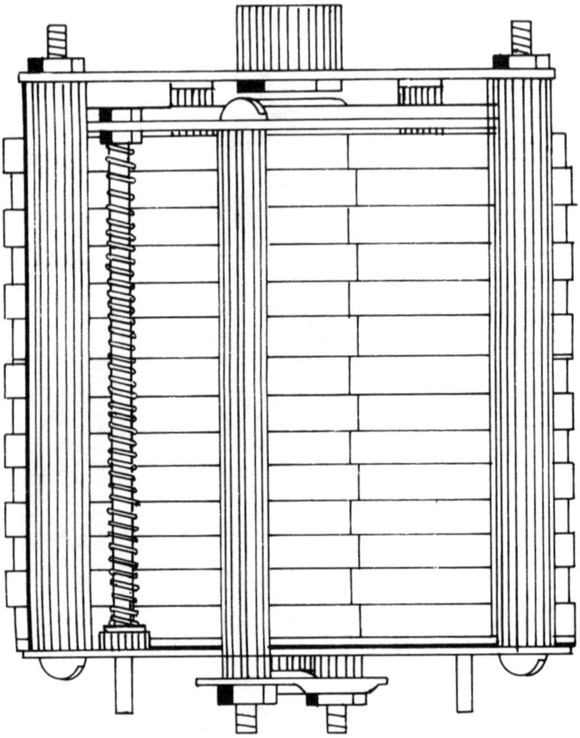

Fig. 3-10. Construction of carbon pile.

unit connects the fixed resistance to the load unit leads only when needed. The carbon pile is a variable resistance consisting of plates of carbon piled one on top of each other. As the plates are compressed by the control knob on the front panel of the tester, the resistance of the pile varies until the desired load is reached.

It is possible to perform charging system tests without a load unit or a carbon pile. This is done by discharging the battery so that it can be used as a load for the charging system. The procedure to discharge the battery is to turn on the headlights for about five minutes, then crank the engine for about 30 seconds with the secondary coil wire grounded. This reduces the battery voltage sufficiently so that the charging system will be forced to deliver full output current.

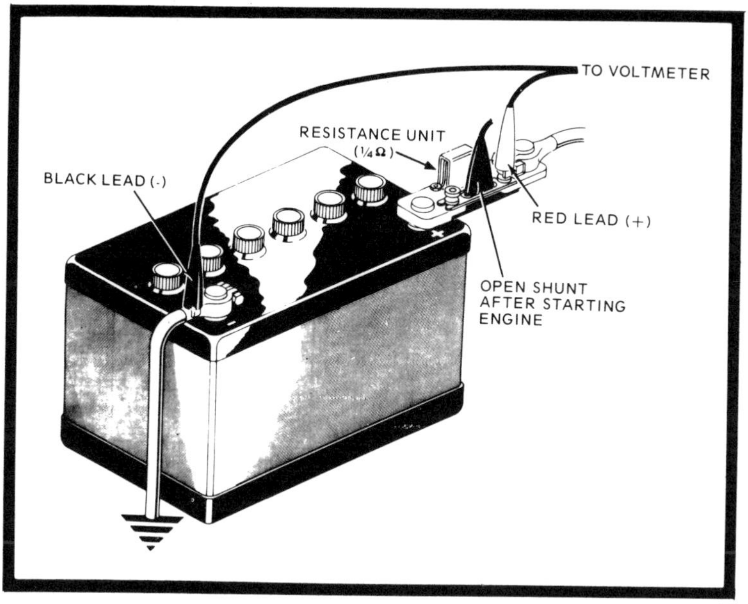

Fig. 3-11. Connections for battery post adapter and voltmeter for voltage regulator test.

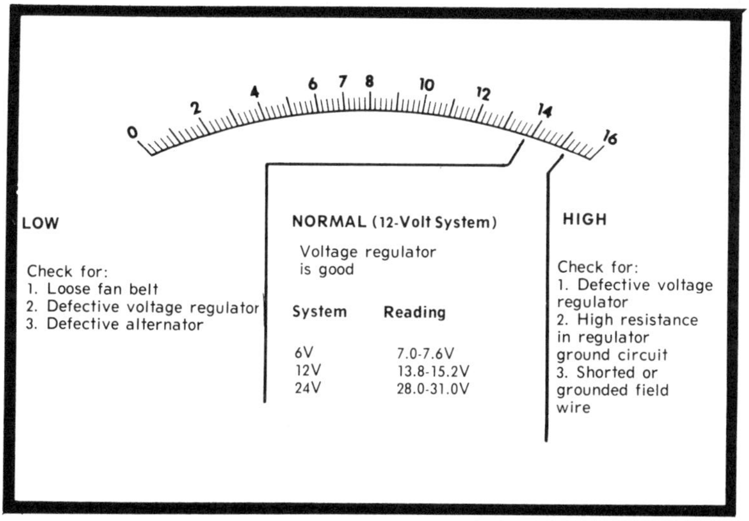

Fig. 3-12. Meter interpretations for voltage regulator test.

ALTERNATOR TESTING PROCEDURE

The procedure for checking an alternator system can best be understood by breaking it down into steps. Four steps are required:

1. Voltage regulator test
2. Current output test
3. Circuit resistance test
4. Diode test

Alternator Voltage Regulator Test

The alternator voltage regulator test is made to determine if the voltage regulator is operating properly within manufacturer's specifications. A regulator setting which is higher than specified can cause battery overcharge and damage to lights and accessories. A lower than specified setting can result in a discharged battery.

The hook-up for this test is shown in Fig. 3-11. The resistance unit (¼ ohm) is required for this test. The 16-volt range on the voltmeter is used.

With the engine running at 1500 rpm, and with the shunt open, the reading of a normal 12-volt system should be between 13.8 and 15.2 volts. A reading either lower or higher than this indicates trouble.

As shown in Fig. 3-12, a high voltage reading ordinarily indicates that the regulator is defective. However, in some instances, a high voltage reading may be caused by high resistance between the case of the regulator and ground. In addition, a high voltage reading may be caused by a field wire shorted to the plug terminal of the alternator.

If the fan belt is tight, a low voltage reading means that the alternator or regulator is defective. To pinpoint the problem, perform the alternator current output test.

Alternator Current Output Test

The alternator current output test is conducted to determine if the alternator has sufficient current output to meet the demands of the vehicle. The hook-up for the alternator current output test is shown in Fig. 3-13. Only the ammeter is needed for this test. The example, shows the use of the load unit which, although not essential, simplifies the test.

Prior to performing the test, the mechanic must look up the current rating of the alternator. The average current rating for a standard system is 30 amps. A system with air conditioning will have a rating of approximately 40 amps.

The test is performed by depressing the switch on the load unit in order to place the load across the battery, and by increasing the engine speed to approximately 2000 rpm. If the reading on the ammeter is within 5 amps of the rated output, the alternator is performing properly.

If, as shown in Fig. 3-14, the rated output current is not obtained, the regulator must be eliminated by using a jumper wire as shown in Fig. 3-15.

On an A circuit, disconnect the lead to the field terminal and connect a jumper wire between the field terminal and ground. On a B circuit, disconnect the lead to the field terminal and connect a jumper wire between the battery and field terminals. If there is doubt as to whether the circuit is A or B, the test may be performed with the jumper tried both ways.

Once the regulator is eliminated, the test is repeated. That is, the load is applied and the engine speed is again increased to 2000 rpm. If, this time, the rated current output is obtained, the alternator is good and the regulator is probably defective. If rated current output is not obtained, the alternator is probably defective.

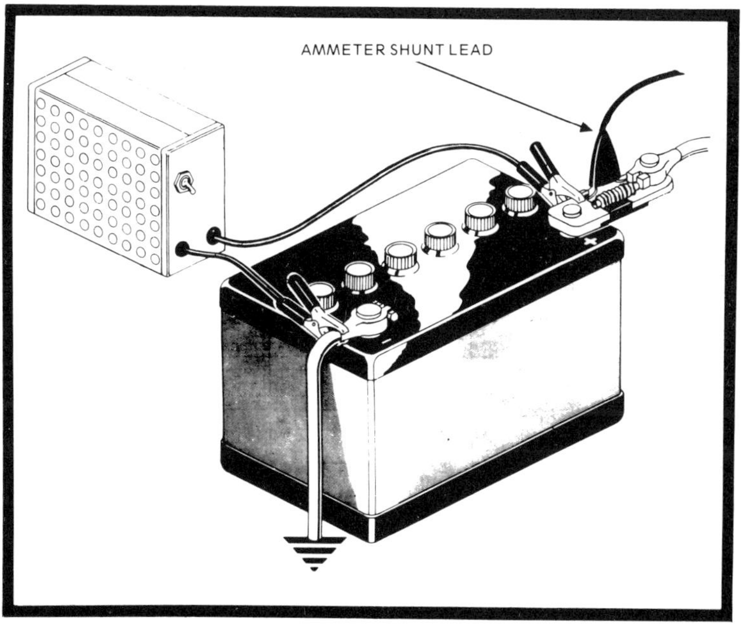

Fig. 3-13. Connections of ammeter and load unit for alternator current output test.

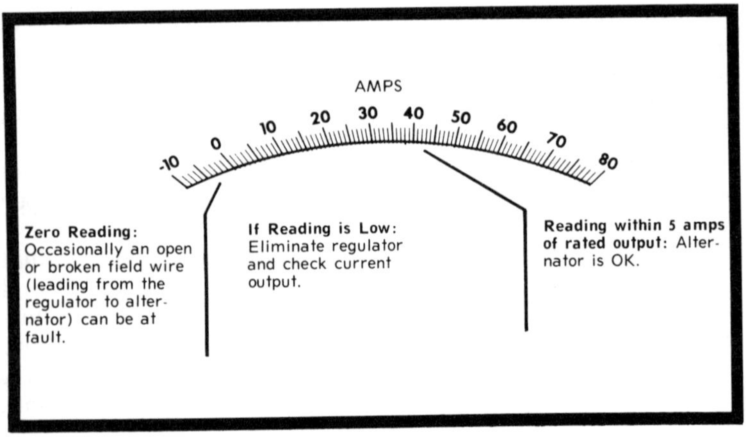

Fig. 3-14. Meter interpretations for current output test.

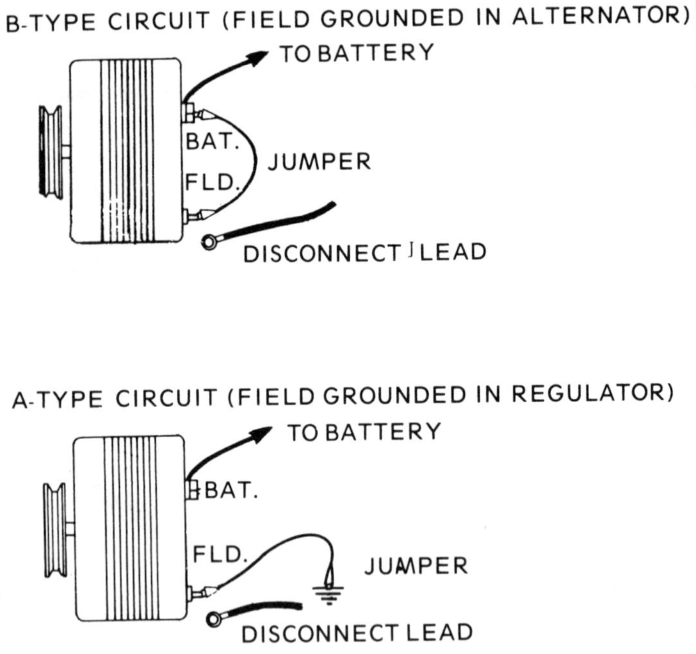

Fig. 3-15. Elimination of regulator with jumper wire.

Alternator Circuit Resistance Tests

The alternator circuit resistance tests are made to determine the amount of voltage loss from the plus battery terminal to the battery terminal on the alternator and from the minus battery terminal to the alternator case. Any voltage loss caused by high resistance in these circuits will reduce overall charge rate and can lead to eventual battery discharge. A high resistance can be present in the form of poor connections or defective or inadequate wiring.

It is important to check that the system is capable of charging before the tests are performed. A load unit should be placed across the battery so as to draw approximately 20 amperes of current from the alternator. The tests are performed by using the low voltage scale on the analyzer. The voltmeter is connected as shown in Fig. 3-16. First it is connected as shown in the A position, then it is connected as shown in the B position. In each position the test is made while the engine is operating at a fast idle (1200 rpm).

The voltage drop read on the voltmeter should be no more than 0.2 volts, as shown in Fig. 3-17. A reading above 0.2 volts indicates excessive resistance which must be corrected.

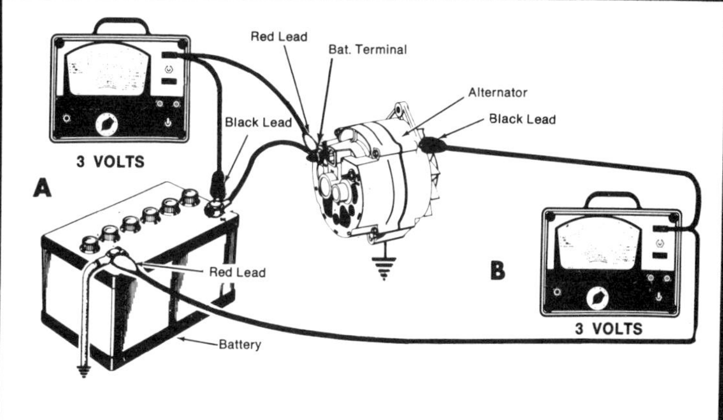

A. IN THE INSULATED CIRCUIT, From the positive battery post to the BAT terminal on the alternator.

B. IN THE GROUND CIRCUIT, From the grounded battery post to the alternator case.

Fig. 3-16. Voltmeter connection for circuit resistance tests.

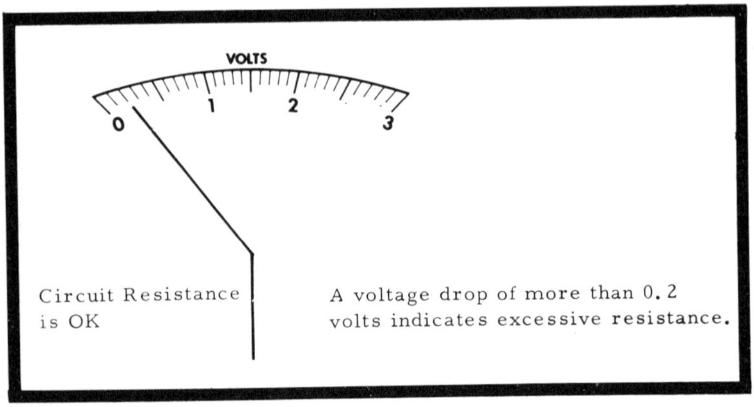

Fig. 3-17. Meter interpretation for circuit resistance tests.

Alternator Diode Tests

Numerous methods exist for testing the diodes in an alternator. The techniques range from the use of an ordinary ohmmeter and voltmeter to an oscilloscope. There are also special purpose testers designed to test only diodes. Here the ohmmeter, voltmeter, and special purpose testers will be discussed. The oscilloscope technique will be explained later.

The low-ohm scale of the ohmmeter should be used when making diode tests. Disconnect the wires going to the diode

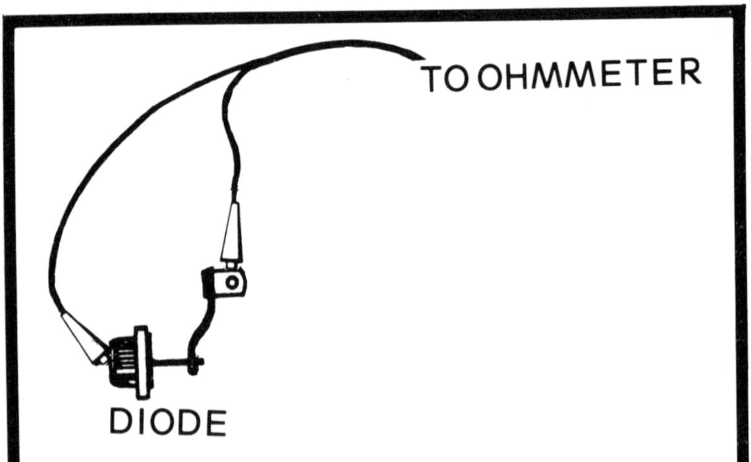

Fig. 3-18. Use of ohmmeter for testing diodes.

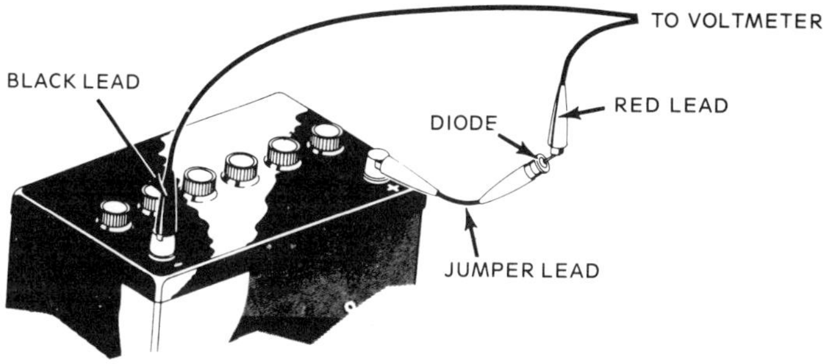

Fig. 3-19. Diode testing with a voltmeter.

lead and connect one ohmmeter test clip to the diode lead. Connect the other test clip to the diode body (Fig. 3-18). Read the meter and then reverse the ohmmeter clips and read the meter again.

With a good diode, the reading with the ohmmeter clips connected one way will be near zero ohms; with the ohmmeter clips reversed, there will be no meter indication (extremely high resistance). With a defective diode, a reading of near zero ohms with the ohmmeter clips connected both ways indicates a shorted diode; no meter indication with the ohmmeter clips connected both ways indicates an open diode.

The diodes may also be tested with a 16-volt voltmeter and a car battery. Using a jumper lead, connect the diode to the car battery as shown in Fig. 3-19 and take a voltage reading. Then turn the diode end for end and take a second reading. The results are tabulated in the table shown in Fig. 3-20.

Fig. 3-21 illustrates a tester designed exclusively for the testing of diodes. It has the ability to check complete sets of diodes or rectifier banks without the necessity of isolating the individual components.

A. Good Diode	— One reading of zero and one reading of battery voltage.
B. Shorted Diode	— Both readings of full battery voltage.
C. Open Diode	— Both readings of zero.
D. Leaky (Partially Shorted Diode)	— One reading of battery voltage and the other above zero.

Fig. 3-20. Test results of diode test with voltmeter.

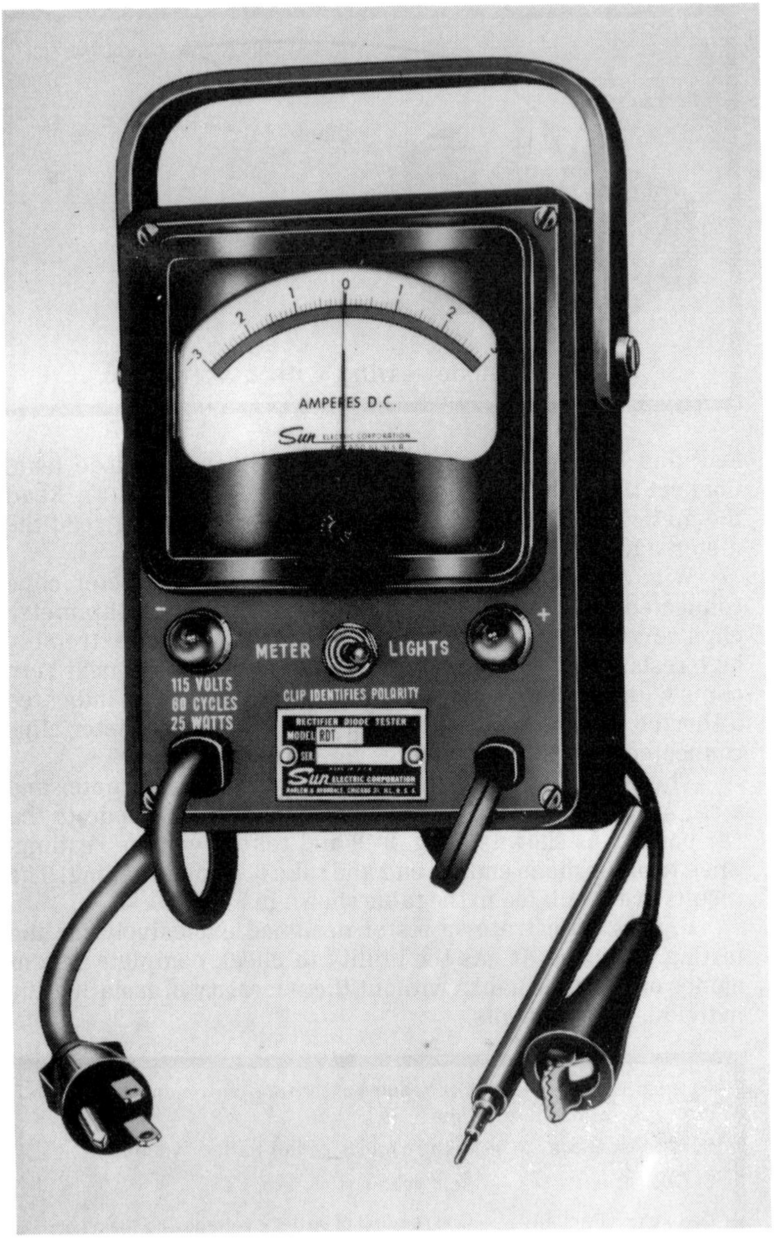

Fig. 3-21. Special purpose tester for diode testing.

The instrument introduces an alternating current into the diode and measures the ability of the diode to rectify the current. In addition, a test light circuit provides a simple reverse current leakage diode test.

Another method of testing the alternator diodes is to measure the ripple voltage while the diode is functioning in the charging system. An instrument that can check diodes this way is shown in Fig. 3-4. A low ripple voltage indicates that the diodes are normal. If the reading is high, one or more shorted or open diodes may be the cause (Fig. 3-22).

GENERATOR TESTING PROCEDURE

The procedure for testing a generator system consists of the following steps:

1. Voltage regulator test
2. Current output test
3. Circuit resistance test
4. Cutout relay test
5. Current regulator test

Generator Voltage Regulator Test

The generator voltage regulator test is very similar to the alternator voltage regulator test. The hook-up is identical to

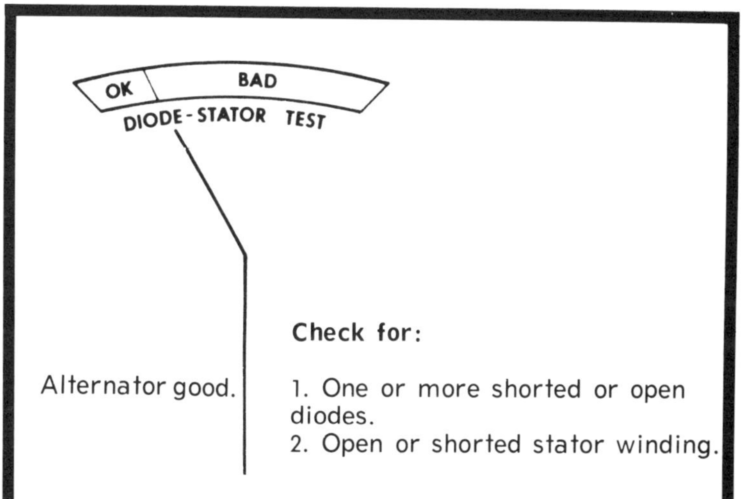

Fig. 3-22. Meter indications for diode test.

SYSTEM	REGULATOR VOLTAGE
6-volt systems	7.0 - 7.6
Chrysler	14.3 - 14.6
General Motors	13.8 - 14.8
Ford	14.8 - 15.6

Fig. 3-23. Typical voltage regulator settings for generator system.

that shown in Fig. 3-11. After the engine is started, the shunt on the battery post adapter is opened, and the engine speed is adjusted to approximately 1200 rpm. The reading of the voltmeter is then compared with the voltage regulator specification. If specifications are not available, typical voltage regulator settings are indicated in Fig. 3-23.

A high voltage reading can be caused by either a defective generator or a faulty regulator. To isolate the problem, the field wire at the generator should be removed. If the voltage stays high, the generator is defective. If the voltage goes down the regulator is defective.

A low voltage reading can be caused by either a defective regulator or generator. In order to pin-point the problem, the generator current output test must be performed.

Generator Current Output Test

The generator current output test is performed to determine if the generator has sufficient current output to meet all the demands required by the car. The hook-up of the tester is similar to the hook-up shown in Fig. 3-13 for the alternator current output test. Engine speed is adjusted to 1600 rpm, and the switch on the load unit is depressed. The test can also be performed without a load unit, as explained earlier, by discharging the battery to the point where full generator current will be called for to recharge it. In either case, the charging current is read on the ammeter. If the reading is within 5 amperes of the rated output current the generator is operating satisfactorily.

If rated output current is not obtained, stop the engine and eliminate the regulator by connecting a jumper wire as shown in Fig. 3-24. Start the engine again and slowly increase speed until rated output current or 2000 rpm is reached. If the current indicated on the meter now rises to within 5 amperes of rated output current at approximately 2000 rpm, the generator is good. A defective regulator or faulty wiring is probably the cause of the trouble.

If the current does not rise to the current specification, the generator is probably defective. The generator circuit resistance test should be performed before any replacements are made.

Generator Circuit Resistance Tests

The circuit resistance tests are made to determine the amount of voltage loss from the generator armature terminal to the positive battery post, the regulator case to ground, and the generator case to the grounded battery post. Excessive voltage loss due to poor connections or defective wiring can cause a reduction in the overall charge rate and can lead to eventual battery discharge.

The test is performed by using a low voltage voltmeter. The voltmeter is connected in the three places indicated in

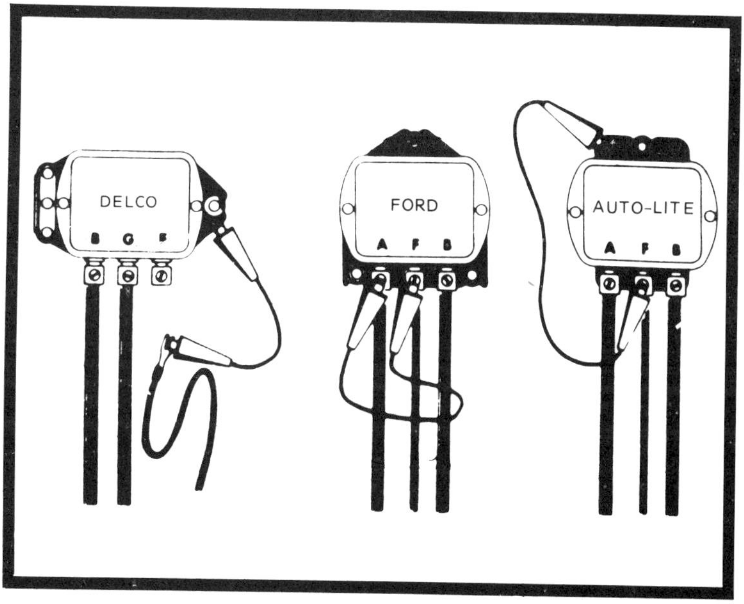

Fig. 3-24. Method of eliminating regulator by using jumper wire.

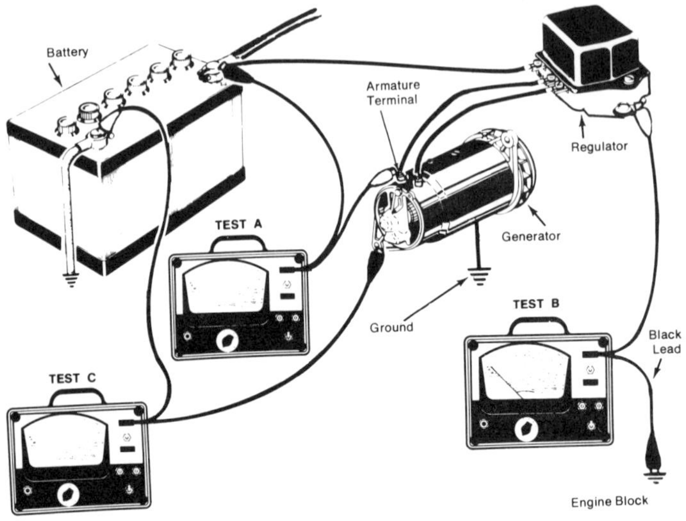

A. IN THE INSULATED CIRCUIT
From the generator armature terminal to the positive battery post.

B. IN THE GROUND CIRCUIT
From the regulator case to the engine block.

C. IN THE GROUND CIRCUIT
From the generator case to the grounded battery post.

Fig. 3-25. Voltmeter connections for circuit resistance tests.

Fig. 3-25. In all three positions the test is performed while the engine is operating at a fast idle (1200 rpm) in order to draw charging current from the charging system.

The voltage drop read on the voltmeter should be no greater than 0.8 volts for the first test and no greater than 0.2 volts for the other two tests. A reading above these specifications indicates excessive resistance which must be corrected.

Generator Cutout Relay Test

The cutout relay test is made to determine whether or not the cutout relay closes and opens properly with respect to generator voltage and battery current. Unless the relay is operating within manufacturer's specifications, a discharged battery or damage to the charging system can result.

Although the cutout relay tests can be performed with a one meter instrument, the better method utilizes two meters. An ammeter and voltmeter are connected as shown in Fig. 3-26.

The engine is operated until the regulator is at operating temperature. Then the engine speed is gradually increased from a slow idle speed. When the points in the regulator close, the ammeter needle deflects in the positive direction to indicate a charging current. The voltage reading at the instant before the points close is the cut-in voltage, sometimes called the cutout relay closing voltage.

Engine speed is then slowly decreased. As the speed decreases, the charging current will decrease. The ammeter needle will move to zero then deflect in the negative direction. When the contacts open, the ammeter needle will again return to zero. The amperage reading just before the contacts open is the reverse current required to open the contacts. It is sometimes called the cutout relay opening current.

The cutout relay is operating satisfactorily when points close and open within the specified limits. When the cutout relay closes and opens at other than specified limits, it should either be repaired or replaced.

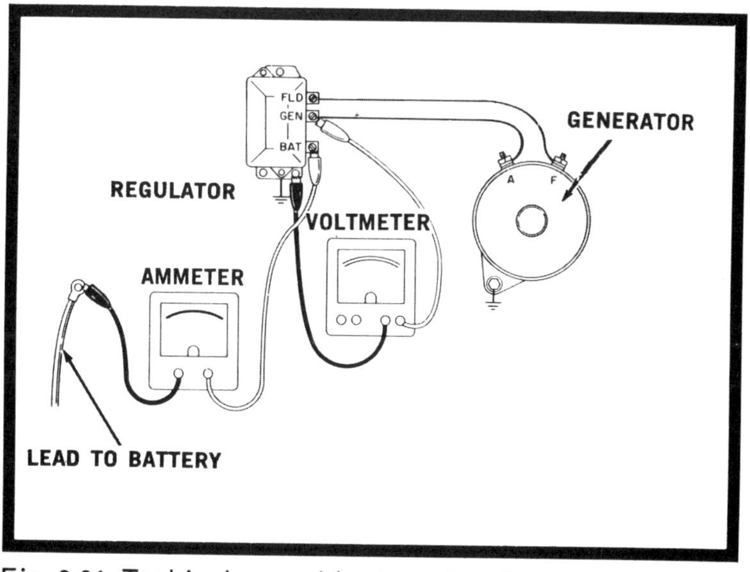

Fig. 3-26. Test instrument hook-up for checking regulator cutout relay.

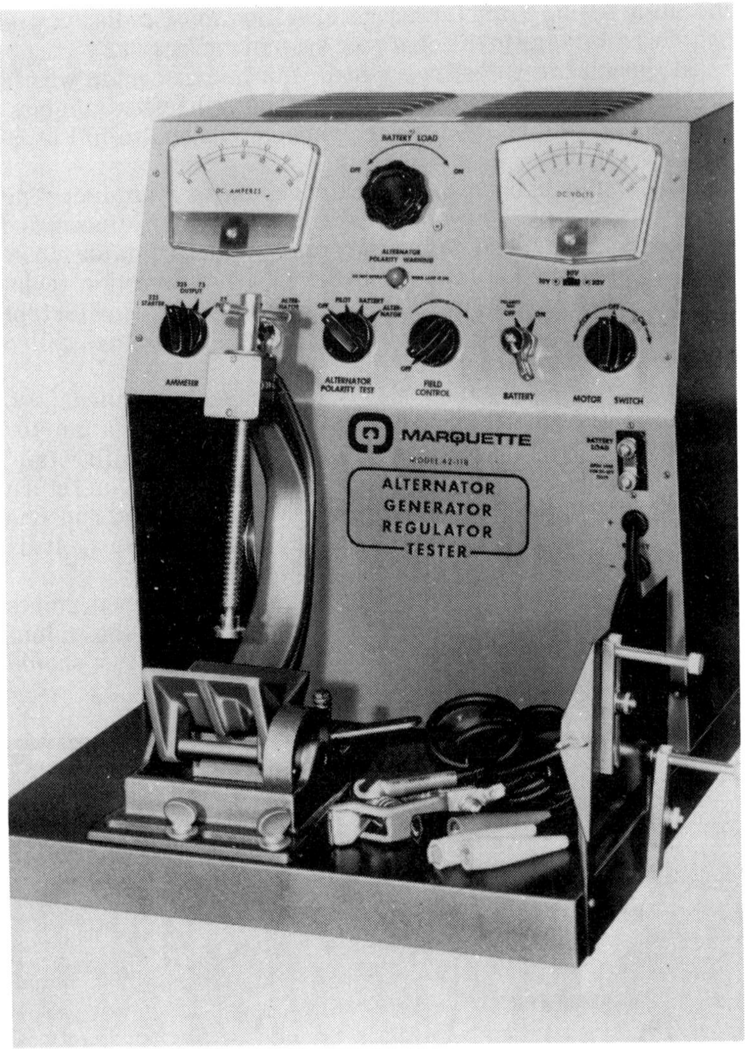

Fig. 3-27. Charging system test bench.

Generator Current Regulator Test

The current regulator test is made to determine if the current regulator is operating properly. A regulator setting which is higher than specified can allow the generator to exceed its rated output; and, consequently, damage from overheating can result. A regulator setting which is lower than specified will not allow the generator to produce the current demanded by the electrical system when loads are great; and, as a result, the battery can become discharged.

The test is performed by connecting an ammeter to measure charging current. The headlights and all accessories are turned on to bring the system voltage down below the voltage regulator setting. The engine is started and speed is increased. The current output reading should increase to the specified current regulator setting and then hold that output as the speed is further increased. If the current is higher or lower than specified, the current regulator is defective or improperly adjusted.

Charging System Test Benches

The test bench illustrated in Fig. 3-27 is designed to simulate the operating conditions of a charging system. The alternator or generator and regulator are tested on the bench so that the mechanic will know that they will perform properly when installed on an automobile. The cost of a test bench ranges from approximately $600.00 to over $8000.00, so every shop is not justified in buying one.

The advantage of a test bench is that it saves time and gives much more positive results than can be obtained without it. It is true that the voltage and current output of a charging system can be measured on a car, but if adjustment of the regulator or work on the alternator or generator is to be done, it is far better to remove the units from the car for testing and adjusting on the test bench.

Chapter 4

Tachometers and Dwell Meters

The tachometer and dwell meter will be considered together. These meters frequently appear together in one piece of test equipment. Such a combined instrument is usually called a tach-dwell tester. Most tach-dwell testers also have a point resistance test. All three of these tests will be examined in this chapter.

TACH-DWELL TESTERS

There are two basic types of tach-dwell instruments. The power to operate the circuit in the less costly instrument, Fig. 4-1, comes from the vehicle's ignition system. This means that the ignition system to which the tester is connected is electrically loaded by the instrument. In addition, abnormalities in the ignition system can result in inaccurate readings on the instrument. Calibration, however, on these instruments is seldom required.

The other type of instrument, Fig. 4-2, while more expensive, is not as subject to inaccuracies due to ignition problems. This type is powered by an external source, either a battery or AC line voltage. In addition, it ordinarily has a set control to adjust the calibration of the instrument for the type of engine on which it is to be used. For all practical purposes, this type of instrument does not draw any power from the ignition system to which it is connected. It can also be used on a greater variety of ignition systems.

Another tach-dwell unit is shown in Fig. 4-3. This instrument uses separate meters for the tach and dwell measurements, but in all other respects, it is similar to the instrument just described.

Tachometer Meter Scales

Engine speed is measured in revolutions per minute (rpm). On the type of tachometer used for automotive service work there is usually a low and a high rpm scale. The ranges

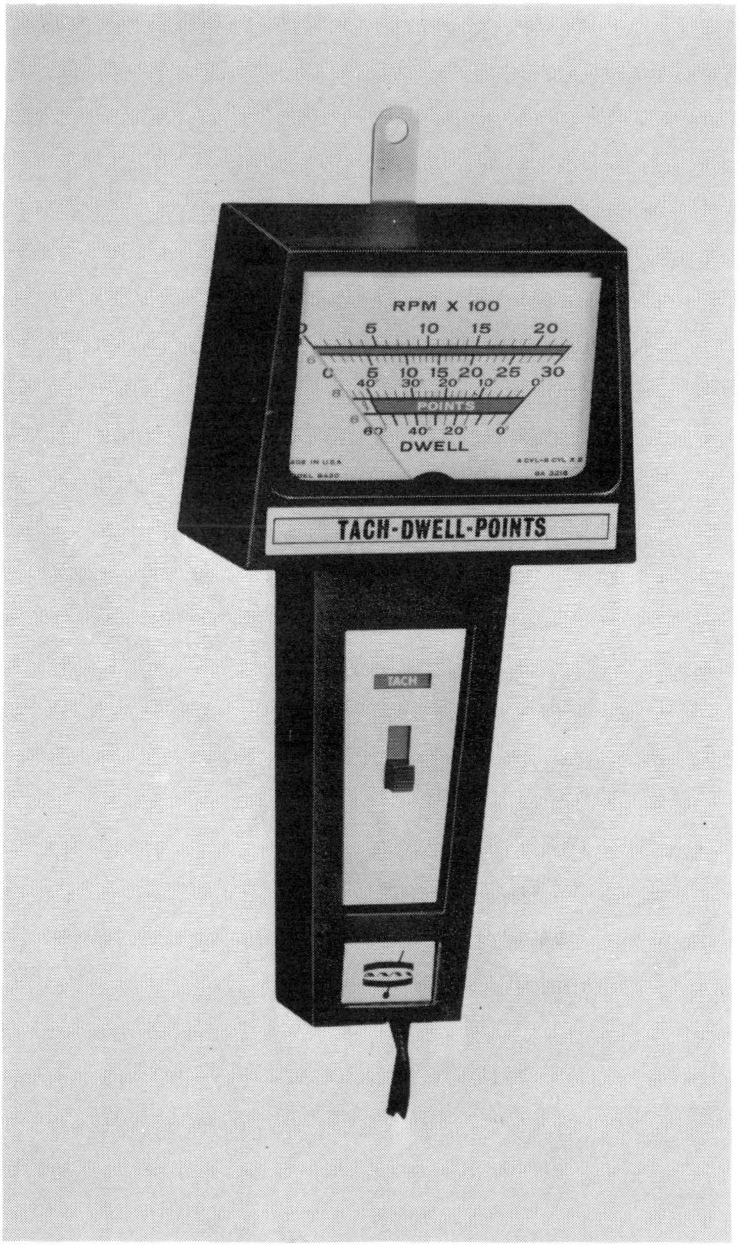

Fig. 4-1. Tach-dwell meter powered by vehicle's ignition system.

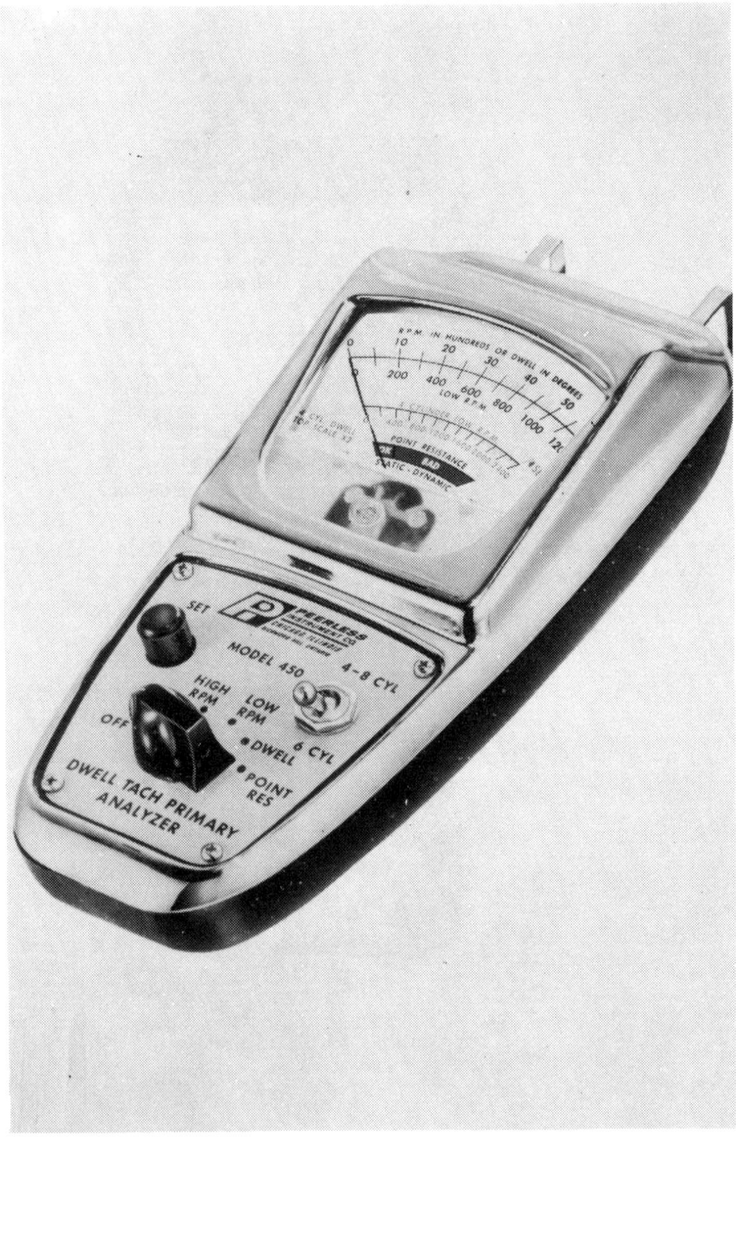

Fig. 4-2. Tach-dwell meter powered by external source.

Fig. 4-3. Two-meter tach-dwell tester.

that are used vary slightly from instrument to instrument as can be seen below:

LOW RANGE	HIGH RANGE
0-900 rpm	0-4500 rpm
0-1000 rpm	0-5000 rpm
0-1200 rpm	0-6000 rpm

Two speed ranges make reading of the scale more convenient when the speed is under 900 or 1200 rpm, as the scale divisions are 5 times larger than the divisions on the high speed scale.

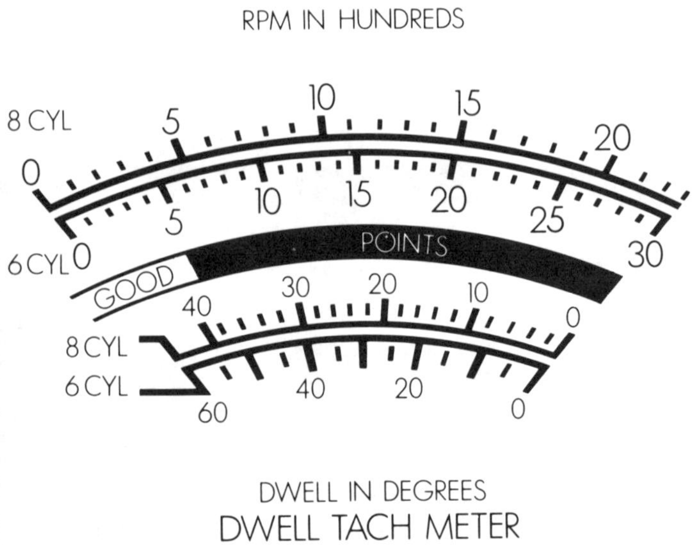

Fig. 4-4. 8- and 6-cylinder tachometer scales.

The rpm shown on a tachometer scale is usually engine rpm. The relationship between engine rpm and distributor rpm for a 4 cycle engine is two to one. This means that for every rotation of the distributor rotor, the engine turns over twice. For example, an engine rpm reading of 2000 rpm is the same as a distributor rpm reading of 1000 rpm.

The difference in the number of cylinders in the engine is accounted for by several methods. One method is to place two scales on the dial, one for 6-cylinder engines and another for 8-cylinder engines, as shown in Fig. 4-4. Nothing is done to set or calibrate the meter. The correct rpm is read on the appropriate scale. For example, one meter indication could be 2000 rpm for a 6-cylinder engine or 1500 rpm for an 8-cylinder engine. Obviously, a 4-cylinder scale could also be placed on the meter. The same meter indication as in the example would then be 3000 rpm (double the 8-cylinder reading).

Another method utilizes one scale for both 6- and 8-cylinder engines, but the operator is required to adjust a set control to a calibration point shown on the meter (Fig. 4-5).

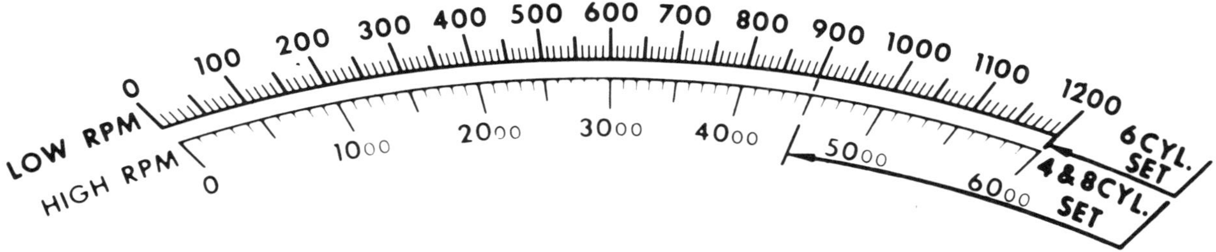

Fig. 4-5. Tachometer scale requiring SET adjustment.

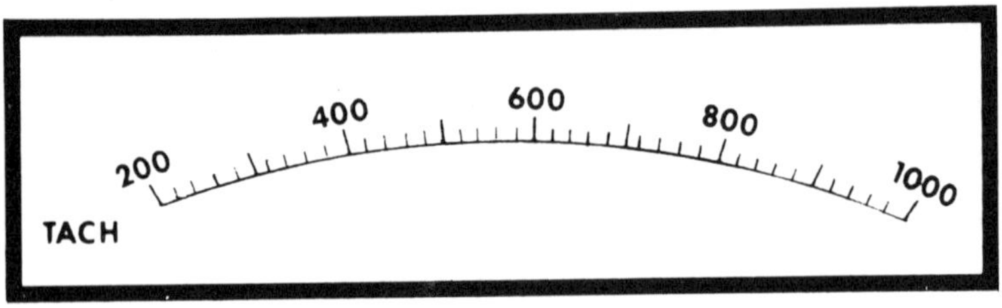

Fig. 4-6. Expanded scale tachometer.

This is done by rotating a switch to a set position; then, by carefully adjusting a set control knob, the meter indication is adjusted to either the 6- or 8-cylinder set point. This adjustment can be made at any time, even when the meter is connected to a running engine. Once the instrument has been set correctly, the scale can be read directly for 6- and 8-cylinder engines. For 4-cylinder engines, the 8-cylinder reading is doubled.

Still another method utilizes one scale for 4-, 6-, and 8-cylinder readings, but in place of the manual set operation described above, a 4-, 6-, and 8-cylinder selector switch on the front panel of the instrument allows the operator to select the correct calibration of the meter.

Since a tachometer is seldom used below 200 rpm, an expanded scale, such as shown in Fig. 4-6, is frequently used for the low rpm range. No reading is indicated by the tachometer until 200 rpm is reached. Beyond 200 rpm the meter responds as an ordinary tachometer; but, since the spacing between the divisions is increased, the accuracy with which the meter can be read is also increased.

Readings for engines other than six or eight cylinders can be made by converting 8-cylinder readings to the desired cylinder reading. The table shown in Fig. 4-7 gives the conversion factors. As an example, if the reading on the 8-cylinder

2-CYCLE ENGINES	4-CYCLE ENGINES
1 Cylinder - Meter Reading times 4	1 Cylinder - Meter Reading times 4
2 Cylinder - Meter Reading times 2	2 Cylinder - Meter Reading times 2
4 Cylinder - Meter Reading times 1	4 Cylinder - Meter Reading times 2

Based on engines without distributor, with two point closures for every revolution.

Fig. 4-7. Table for converting 8-cylinder meter readings to other cylinder readings.

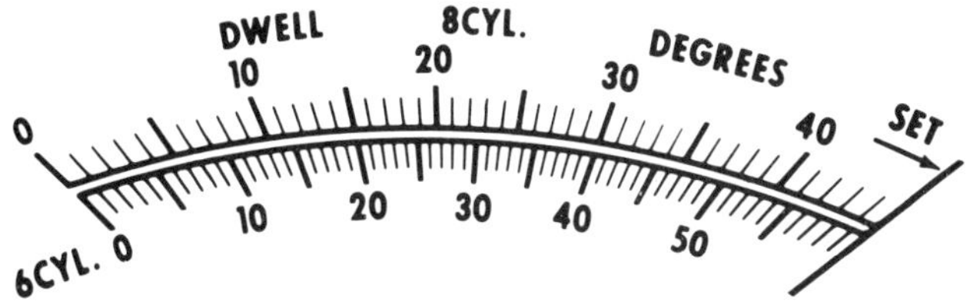

Fig. 4-8. Forward reading dwell scale.

scale is 550 rpm, the correct speed reading for a 4-cylinder 4-cycle engine is 1100 rpm.

When making conversions from one scale to another, it is important to remember the distinction between a 2- and 4-cycle engine. The usual type of automotive cycle is the four-stroke cycle, delivering a power pulse to the crankshaft once for every two revolutions of the crankshaft. The two-stroke cycle delivers a power pulse to the crankshaft for each revolution of the crankshaft.

Dwell Scales

Dwell scales are either forward reading or reverse reading. The forward reading dwell scale, Fig. 4-8, is found on the more expensive dwell meters, while the reverse reading dwell scale, Fig. 4-9, is found on the less costly instruments.

On dwell meters that do not have a cylinder selector switch, a separate 6- and 8- and sometimes a 4-cylinder scale will appear on the dial. If the separate 4-cylinder scale is omitted, readings can be taken on the 8-cylinder scale and multiplied by two. For example, a 20-degree reading on the 8-

Fig. 4-9. Reverse reading dwell scale.

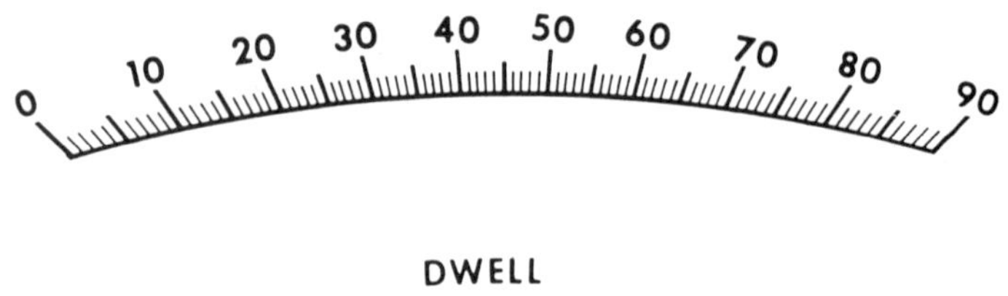

Fig. 4-10. Single scale for all cylinder engines.

cylinder scale is a 40-degree dwell reading for a 4-cylinder engine.

Instruments provided with a cylinder selector switch have only one dwell scale as shown in Fig. 4-10. All dwell readings, regardless of the number of cylinders in the engine, are made directly on this scale.

Hook-Up of Tach-Dwell Tester

A very desirable feature of the tach-dwell tester is the simplicity with which it may be connected to an engine, see Fig. 4-11.

One wire, usually coded red or yellow, is clipped to the distributor side of the coil or to the insulated terminal on the distributor. The other wire, usually coded black, is connected to a good ground. On a positive ground system the leads are reversed. That is, the black lead is connected to the distributor side of the coil and the red or yellow lead is connected to a good ground. If the leads are connected incorrectly, there will either be no reading or the reading will be quite erratic. This same simple lead connection is used for making tach, dwell and point resistance tests.

Low rpm Tach Test

One of the important adjustments to be made, during a tune-up or when the carburetor is being adjusted, is idle speed.

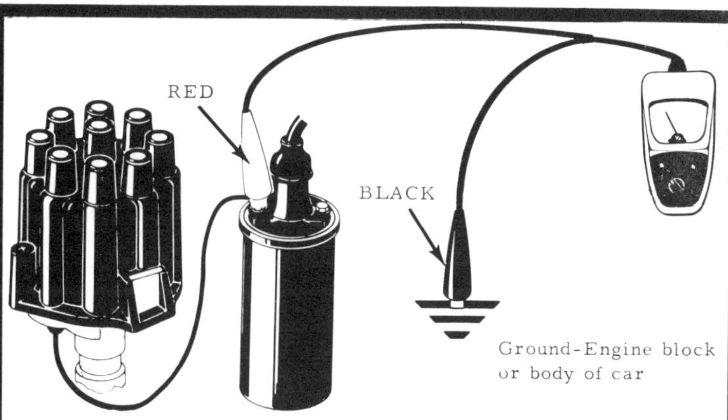

Fig. 4-11. Hook-up to tach-dwell meter.

On cars equipped with a conventional 3-speed transmission, the idle speed is more a matter of preference than anything else, as it does not matter much whether the speed is 400 rpm or 600 rpm, as long as the engine does not stall or is not too noisy as it idles. But, on automobiles with automatic transmissions, the idle speed is quite important. It must be fast enough to keep the engine from stalling when the car makes traffic stops, yet slow enough to prevent excessive creeping when the transmission is in DRIVE and the car is supposed to be standing still. The correct idle speed must be obtained from the manufacturer's specifications. The test is performed by connecting and calibrating the instrument as previously discussed and reading the speed on the appropriate scale (Fig. 4-12).

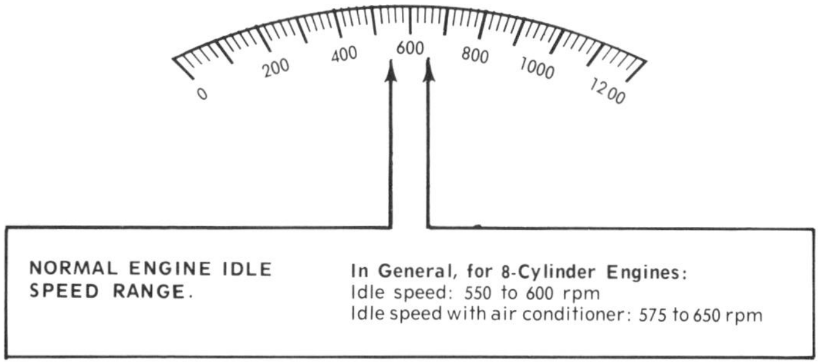

Fig. 4-12. Meter reading for low rpm tach test.

High rpm Tach Tests

The high rpm scale is used for setting governors and adjusting high speed mixture screws on industrial engines. In addition, this scale is frequently required when making tests with other test instruments. For example, when checking the distributor automatic advance, the speed at which specified angles of advance should occur, must be known.

Also, there are charging system and ignition tests that are performed at speeds such as 1500 rpm or 2000 rpm. These various applications, all of which are important, make the tachometer one of the most useful instruments in any shop where tune-up or troubleshooting is done.

Point Resistance Test

Most tach-dwell testers have a point resistance test which can be either a static or dynamic test. The test is a measurement of the voltage drop across the ignition points from the distributor side of the coil to ground. If the reading is within the limits, the point resistance is satisfactory. If it is outside the limits, there is excessive point resistance which should be corrected.

The static point resistance test is performed with the points closed and the ignition switch on. This allows current to flow in the primary circuit so that a voltage drop will appear across the points. The instrument is connected in the normal manner from the distributor side of the coil to ground, and the meter indications are interpreted as shown in Fig. 4-13. The test can also be performed with ordinary voltmeter. A voltage reading of 0.2 volts or more is considered to be excessive and should be corrected.

The dynamic point resistance test can be performed while the engine is running. This eliminates the time it takes to close the points, which usually requires bumping the engine with the cranking motor several times until they close. The results of this test is exactly the same as indicated in Fig. 4-13. When testing a double point system, one set of points should be blocked open while the other set is being tested.

Dwell Angle and Dwell Test

The dwell angle (or cam angle, as it is sometimes called) is the angle through which the distributor shaft turns, while the distributor breaker or contact points are closed and current is able to flow through the primary ignition circuit.

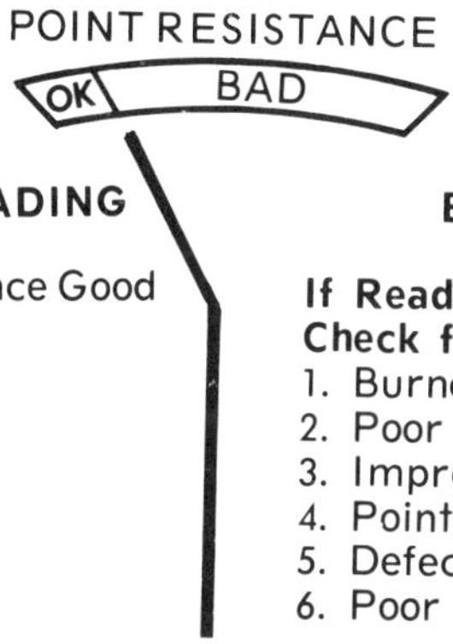

GOOD READING

Point Resistance Good

BAD READING

If Reading is Bad or Erratic Check for:

1. Burned ignition points
2. Poor lead connections
3. Improper point installation
4. Points misaligned
5. Defective ground pigtail
6. Poor distributor ground

Fig. 4-13. Meter indication for point resistance test.

The angle from cam point to cam point for various engines is as follows: for 4-cylinder engines it is 90 degrees, for 6-cylinder engines it is 60 degrees, and for 8-cylinder engines it is 45 degrees. The dwell is always less than these angles, because a complete cycle of point make-and-break must occur during the very brief interval of time that the distributor cam is passing from cam point to cam point. If the points do not break the primary circuit during this time, there is no ignition spark.

The specification for dwell varies considerably for the various engines, so the manufacturer's specifications should be checked before making a dwell test. In the absence of manufacturer's specifications, the table shown in Fig. 4-14 can be used as a general guide. It is essential that this angle be within the specified limits, because if the angle is too small, there will not be sufficient time allowed for the magnetic field

CYLINDER	DWELL ANGLE
8	28°-32°
6	35°-38°
4	50°-60°

Fig. 4-14. Dwell angle for various cylinder engines.

to be built up in the ignition coil, especially at high speed. This will cause the top car speed to be limited, and possibly make the engine misfire. If the angle is too large, the points will not open as far as they should, which frequently results in engine starting difficulty; breaker point burning may also occur.

Ignition timing is also affected by the dwell angle. With a reduced angle, the opening of the points occurs earlier than it should, resulting in excessive ignition advance. On the other hand, excessive dwell causes a retarded spark. Therefore, any time that the ignition timing is to be adjusted, the dwell angle must be checked first to see that it is within the specified range. If it is not, it must be set correctly before making a change in the distributor timing setting. There is always the possibility that someone may have set the timing without first checking the dwell angle; and then, when the dwell angle is set correctly, the timing will be incorrect.

The dwell test is performed by connecting and calibrating the instrument as previously discussed and reading the correct scale while the engine is operating at idle speed, see Fig. 4-15.

Adjusting Dwell

On some distributors (the external adjustment type), the dwell angle can be adjusted with the engine running. On the other distributors (the internal adjustment type), the dwell angle must be set while the engine is cranking. In either case, disconnect and plug the vacuum advance line from the distributor; since, on some distributors, the dwell angle changes with the operation of the vacuum advance. Also, if the vehicle is equipped with an advance-retard solenoid, disconnect the wire at the carburetor end. The procedure for adjusting dwell for both types of systems is as follows:

A. External Adjustment Type (Fig. 4-16)

1. With engine at idle and dwell meter connected, raise window and insert Allen wrench into adjusting screw.

2. Turn adjusting screw, set and read specified dwell on the meter.

B. Internal Adjustment Type (Figs. 4-17 and 4-18)

1. Remove distributor cap and rotor and ground the secondary coil wire.

2. Turn the engine ignition switch on and check the dwell reading while cranking the engine. It is frequently normal for the meter pointer to fluctuate slightly at low engine speed in which case an average reading must be taken.

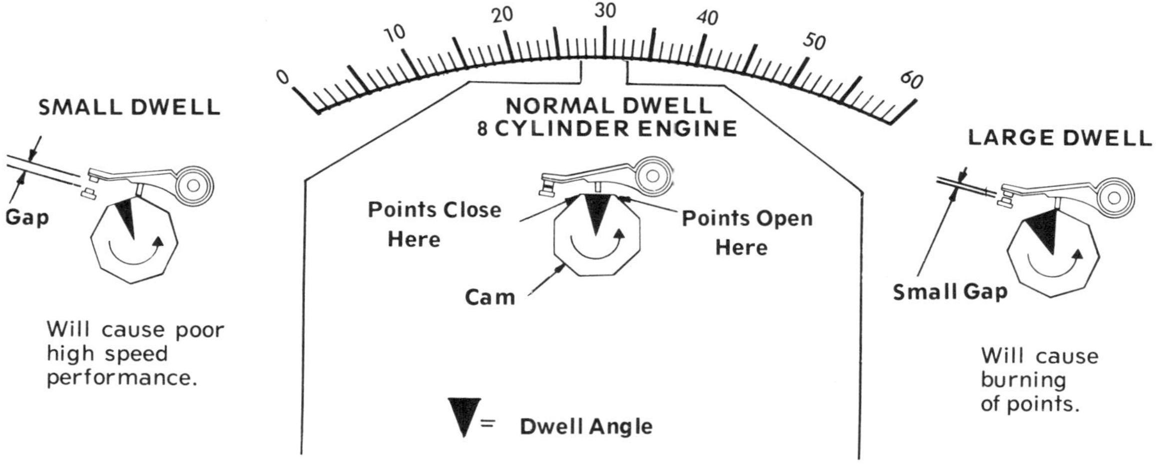

Fig. 4-15. Meter reading for dwell test.

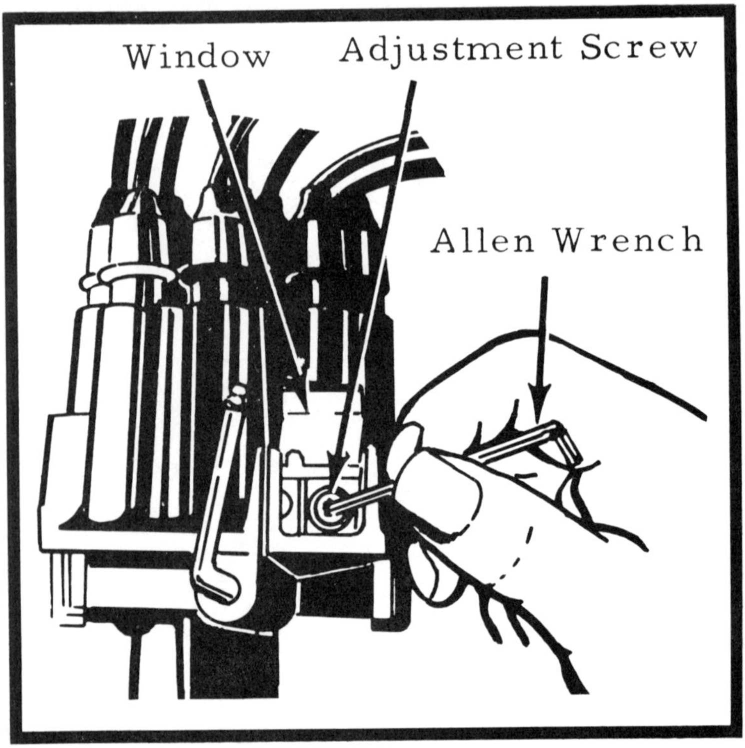

Fig. 4-16. Adjustment of dwell on external adjustment type distributor.

3. To adjust dwell, loosen locking screw slightly and adjust gap by turning adjusting screw (Fig. 4-17) or by inserting a screwdriver in the slotted hole (Fig. 4-18) and turning the tool slightly left or right to obtain the specified dwell. Tighten locking screw and recheck dwell.

4. Reassemble distributor and recheck dwell reading with engine operating at idle speed and with vacuum line disconnected from diaphragm of distributor.

Remember that it is important to recheck the ignition timing every time the breaker points are adjusted. A change of one degree in dwell causes a one degree change in timing.

Dwell Variation Test

When checking the dwell angle with a dwell meter it is advisable to check for variation in dwell at gradually in-

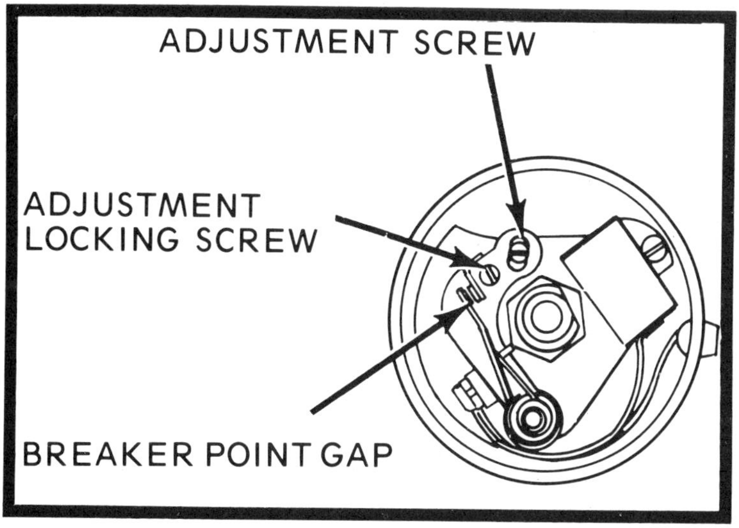

Fig. 4-17. Adjustment of dwell on internal adjustment type distributor by means of turning adjusting screw.

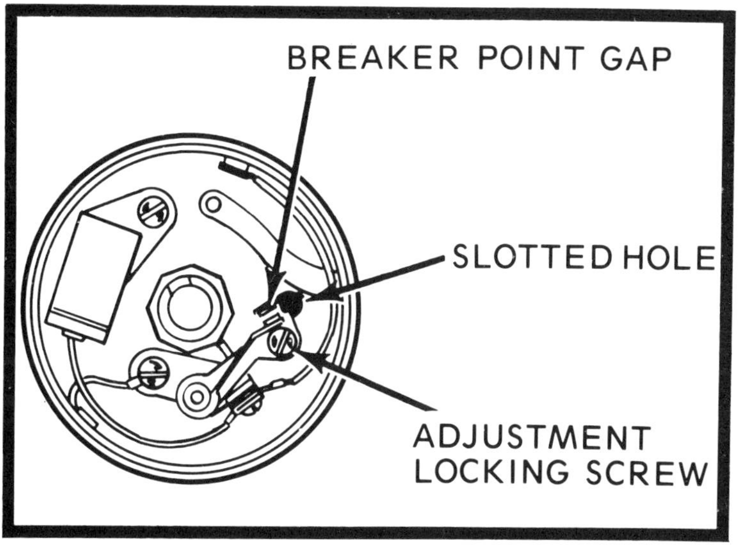

Fig. 4-18. Adjustment of dwell on internal adjustment type distributor by means of rotating screwdriver in slotted hole.

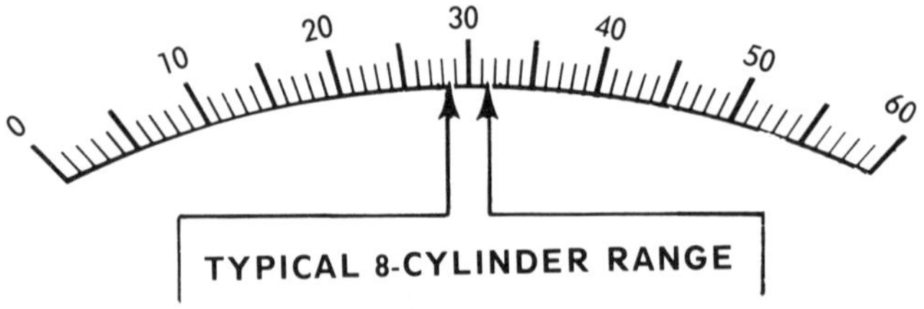

If Variation is More than 3 Degrees, Check for:

1. Wear in distributor shaft and bushings.
2. Wear in breaker plate.

Fig. 4-19. Meter indication for dwell variation test.

creased engine speeds up to 1500 rpm. If dwell varies more than three degrees, as engine speed is increased, the distributor is defective, see Fig. 4-19. The test should be performed with the vacuum advance line disconnected from the distributor, as on some distributors the dwell angle changes with the operation of the vacuum advance. Excessive dwell variation is caused by improper contact spring tension, horn or wobbly breaker plate, worn distributor shaft bushings, or loose stationary contact.

OTHER IGNITION SYSTEMS

There is a great variety of transistorized ignition systems. It would not be possible nor practical to list them all. Fortunately, the systems are basically the same in that the heavy primary coil current is carried by a transistor unit, instead of the distributor points. A typical system and lead connection is shown in Fig. 4-20.

The tachometer can be used on most magneto systems where a connection can be made to the breaker points. On some engines, the hot lead is brought out to the primary shorting switch used to stop the engine. (Not the grounding strap to the spark plug.)

Some capacitive discharge systems are provided with a lug to which a tachometer can be connected. On other systems, the tachometer can usually be connected from ground to the primary wire leading to the breaker points or, in some cases, from ground to the coil primary.

OTHER TYPES OF TACHS

The tachometers that have been considered thus far as those that are commonly known as tune-up tachs. There are several other types of tachometers which are occasionally used in automotive work that should be mentioned. Each one of these has a special purpose. These tachometers are listed below:

A. Secondary Tach
B. Road Tach
C. Phototach
D. Magnetic Tach
E. Strobotach

Secondary Tach

A secondary tach, Fig. 4-21, as its name implies, is connected to the secondary side of the ignition system, instead of to the primary side. This connection may be made either directly to a plug with a trigger transformer, or it may be made by means of a clamp-on pickup that snaps around the spark plug wire. Although a secondary tach can be used on an automotive engine, its greatest use is with small engines on which it is virtually impossible to use an ordinary tach.

In addition, the secondary tach will read correctly for a 4-cycle engine regardless of whether the engine has 2, 4, 6 or 8 cylinders. The secondary tach is energized only when one plug

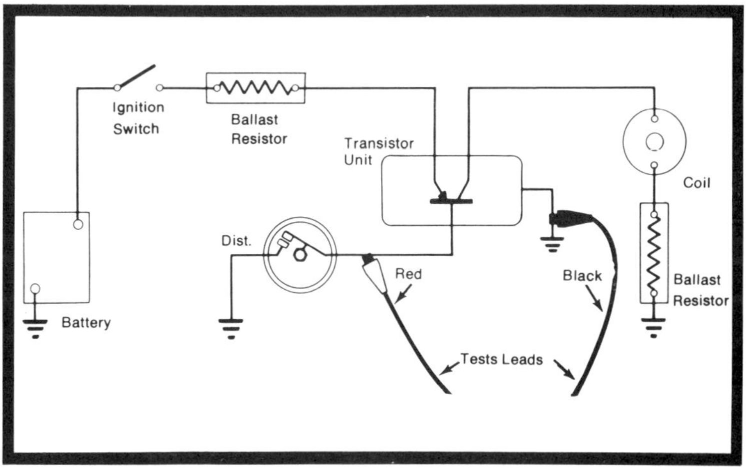

Fig. 4-20. Transistorized ignition system with tachometer connections.

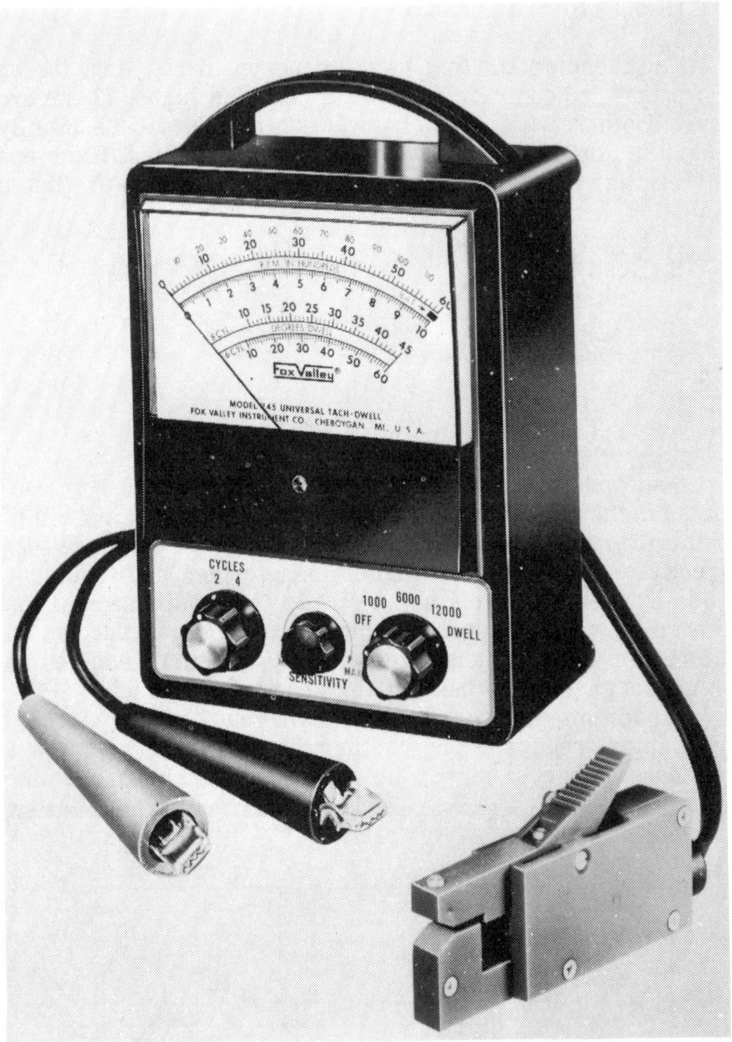

Fig. 4-21. Typical secondary tachometer.

fires, whereas a primary tach is energized every time any plug fires. Therefore, although it is necessary to set a primary tach for the correct number of cylinders, it is not necessary to do so with a secondary tach. The same holds true for a 2-cycle engine except that a separate 2-cycle scale must be used.

Road Tach

A road tach (Fig. 4-22) is designed to be mounted inside the passenger compartment and is used as an aid in driving. A stick-shfit transmission functions at its best when gears are changed at the correct speeds, and the road tach is used to determine when the correct speeds are reached.

Most road tachs have a red-line to simplify their use. The red line is on the face of the dial at the speed at which the gears should be changed. Then, rather than observing the tach readings in rpm, the operator can watch for the meter pointer to align with the set red-line. This is much simpler for the driver to do while trying to watch the road and tach at the same time. The range of a road tach is between 0 and 4000 to 10,000 rpm.

Phototach

A phototach (Fig. 4-23) is a touch-less tach that operates on the principle of reflected light. In order to use a phototach

Fig. 4-22. Example of a road tach.

Fig. 4-23. Typical phototach.

there must be a contrasting surface on the rotating shaft that is to be measured. This can be accomplished by placing a piece of black tape on a highly reflective shaft or a piece of bright tape on a dark shaft. The phototach pickup consists of a light source, similar to a flashlight, and a photocell that detects the reflected light bouncing off the rotating shaft. Each time the photocell detects a pulse of reflected light the tachometer receives a pulse of power. The more reflected pulses the photocell detects the higher the tachometer will read.

The operator must be careful when using a phototach that extra reflected light beams do not reach the photocell. This can occur when there are nicks or cuts in the rotating shaft that create reflective spots.

Magnetic Tach

A magnetic tach is used by placing a magnetic pickup close to a rotating gear as shown in Fig. 4-24. Inside the magnetic pickup, a coil is would around an iron pole piece which is attached to a permanent magnet. When a ferrous object passes closely by the pole piece, the magnetic flux linking the coil increases, and, when the object is removed, decreases. In this manner, passing gear teeth cause the pickup to generate a voltage having a frequency exactly proportional to the number of discontinuities per second interrupting the magnetic field around the pole piece. The voltage drives a tach circuit which is calibrated in rpm.

Typical spacing between the pickup and the rotating gear is 0.005 inches. Consequently, care must be exercised in adjusting the pickup properly before a measurement is made.

Strobotach

A strobotach (Fig. 4-25) is similar to a timing light in that it consists of a flashing light that is aimed at a rotating shaft. But, whereas a timing light is synchronized to engine rpm because its trigger signal is obtained from a spark plug, a strobotach is free running. The strobotach has an oscillator that allows the operator to flash the light at any rpm he selects.

To use a strobotach, the flash of the light is aimed at a rotating object which has been marked with chalk or a piece of tape. The oscillator in the strobotach is then adjusted until the shaft appears to be standing still. At this setting of the

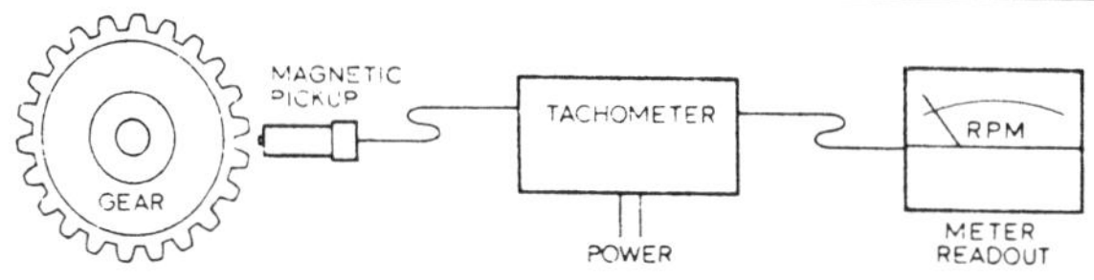

Fig. 4-24. Example of the use of a magnetic tach.

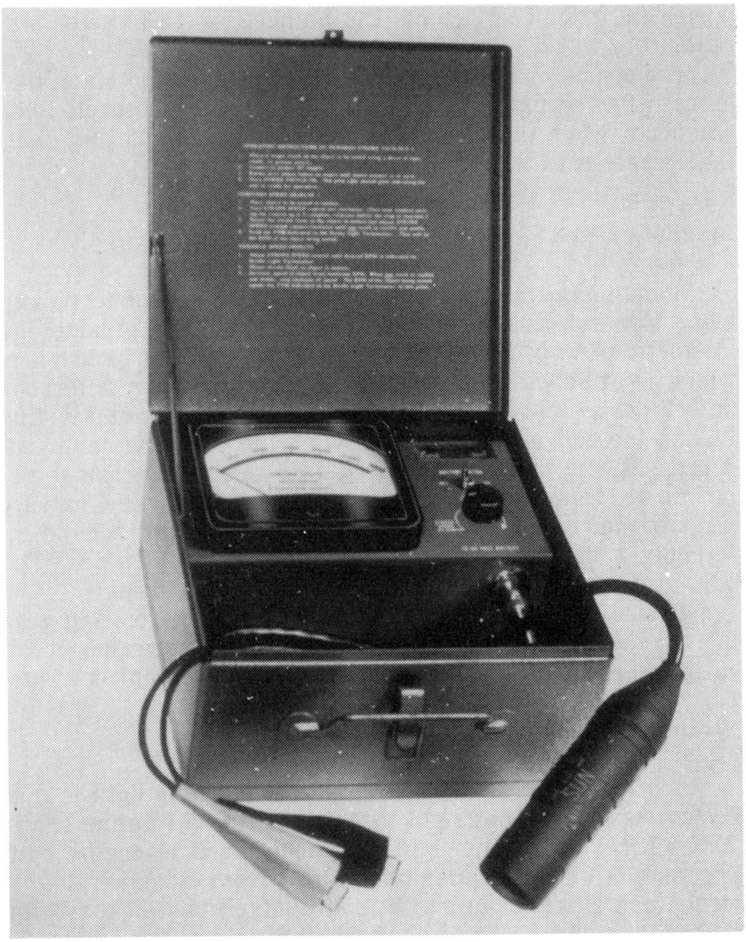

Fig. 4-25. Typical strobotach.

oscillator, the meter will read the correc rpm of the rotating shaft under test.

Care must be taken not to set the oscillator at any multiple of the shaft rpm, because multiple images will also appear to be standing still, and the meter will give an incorrect reading. The rule for setting the strobotach correctly is to adjust the oscillator initially to a speed higher than the speed of the rotating object to be measured. Then, slowly reduce the oscillator speed until the first single image is observed. At this point, the flashing rate of the strobotach will be equal to the rotational speed of the object, and the speed can be read directly from the rpm dial.

Chapter 5

Timing Lights

Proper ignition timing is one of the most important factors relative to the efficient and economical operation of an engine. Improper ignition timing may cause overheating, loss of power, poor acceleration and performance, and may even shorten the service life of the engine.

Ignition timing is ordinarily checked and set with the aid of a power timing light. The light is energized by a power source and triggered by the voltage at the spark plug to which it is connected, usually the No. 1 cylinder. The timing light enables the mechanic to see how the actual timing compares with the specified timing while the engine is operating. If the timing is correct, nothing further needs to be done about the initial ignition timing.

TYPES OF LIGHTS

There are two general types of timing lights—the neon light and the power timing light, which uses a xenon flash tube. The neon light (Fig. 5-1) uses a neon flash tube and is actuated solely by the high voltage at the plug to which it is connected. When it fires it gives a red flash. The power timing light (Fig. 5-2) uses an AC or DC power source and emits a brilliant white or blue-white flash from its xenon flash tube. The AC power source is ordinary line voltage (110 volts) while the DC source is the battery in the car under test.

The voltage on the flash tube in a power timing light varies between 300 and 600 DC volts. This voltage, in the AC timing light, is obtained by means of a voltage doubler circuit connected to the line voltage. In the DC timing light, a vibrator power supply or a transistorized converter is used to obtain a voltage of 600 volts. This higher voltage is required so that a trigger pulse will ionize the gas within the tube and cause it to flash. In all cases the bulb is triggered by means of a high voltage pulse that is obtained directly from No. 1 spark plug or synchronized with the voltage at No. 1 spark plug.

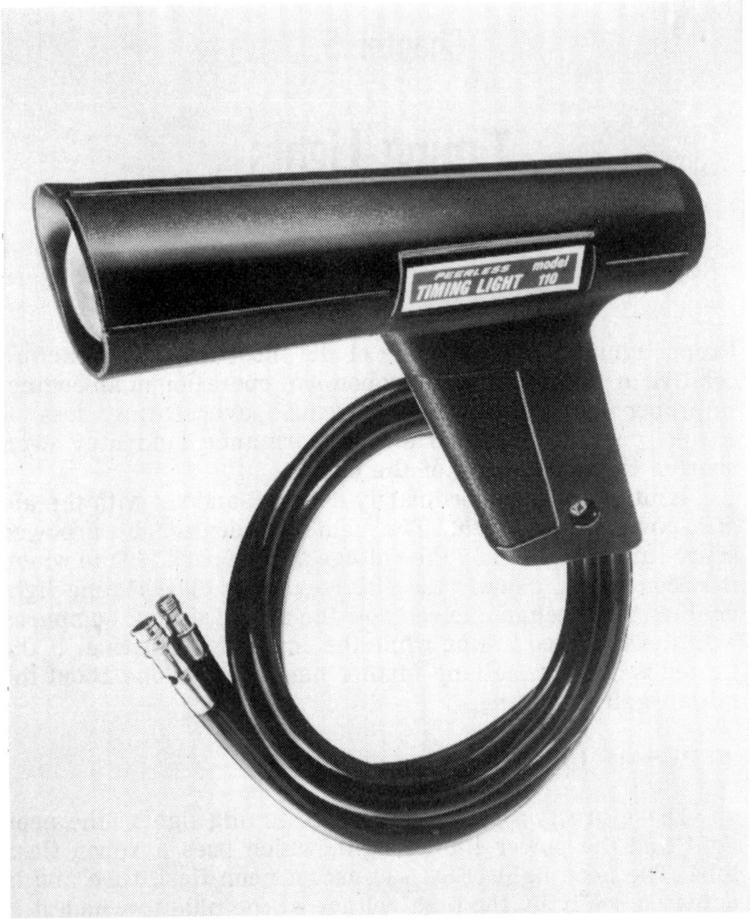

Fig. 5-1. Neon timing light.

Although a timing light is designed for making initial timing checks at idle speed, most timing lights will function up to a speed of 5000 rpm or more. A push button on-off switch is provided on most power timing lights to operate the light only when needed, thereby extending the life of the flash tube.

HOOK-UP

The simple neon light is connected in one of two ways depending on the light itself. Instructions that come with the light will tell whether the high tension lead is connected to No. 1 spark plug with the ignition cable undisturbed, or to the ignition cable disconnected from the plug and attached to one of the timing light cables, with the other cable attached to the spark plug (Fig. 5-3). The latter method puts the light in series with the plug. Whether the car battery voltage is 6 or 12 volts does not make any difference to this type of light, as it works on the high tension side of the ignition system only.

Fig. 5-4 shows how to hook up an AC power light and Fig. 5-5 shows how to hook up a battery-operated power light. Both the AC and battery-operated power lights give enough illumination to permit timing in well lighted garages or even in bright sunlight. The inexpensive neon lights give off a very dim light which can be used to time an engine if bright light is shaded from the area.

Always use the proper adapter when connecting a timing light—never puncture the spark plug cables. In the event the

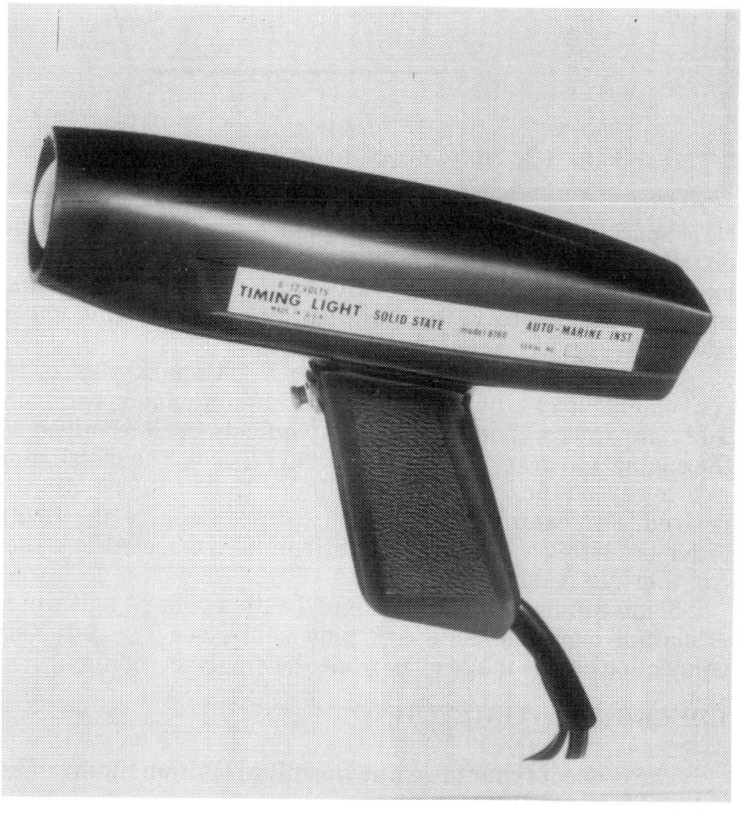

Fig. 5-2. Power timing light.

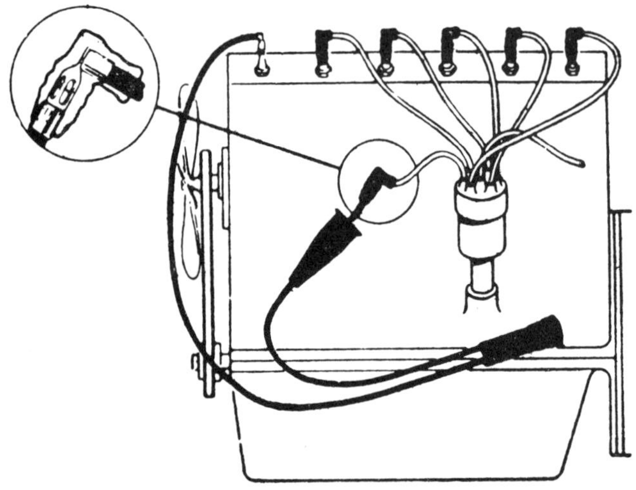

Fig. 5-3. Hook-up of neon timing light.

No. 1 spark plug is not readily accessible, the No. 1 tower in the distributor cap may be used with the aid of an adapter. The location of No. 1 spark plug varies from engine to engine and must be obtained from the specifications of the engine under test.

If connecting the timing light to No. 1 spark plug is inconvenient, it can be connected to the companion cylinder. The companion cylinder to No. 1 can quickly be determined by the engine's firing order as is shown in Fig. 5-6. The distributor cap tower of the companion cylinder may also be used if desired. For example, if on an 8 cylinder engine the firing order is 18436572, the timing light can be connected to either No. 1 or No. 6 spark plug.

Some timing lights are designed with a pick-up unit which is clamped around the spark plug cable, see Fig. 5-7. This connection can be made even while the engine is running.

CHECKING INITIAL TIMING

The procedure for checking the initial ignition timing of an engine is as follows:

1. Locate the timing marks on the crankshaft pulley, harmonic damper, or flywheel, see Fig. 5-8. If they are not

readily visible, wipe them with a cloth and mark them with chalk or paint.

2. Operate the engine until normal temperature is reached.

3. Stop the engine and connect a tachometer and timing light.

4. Disconnect the distributor vacuum line and tape the manifold opening, see Fig. 5-9. This operation is recommended to insure that there is no advance whatever due to vacuum while timing the engine.

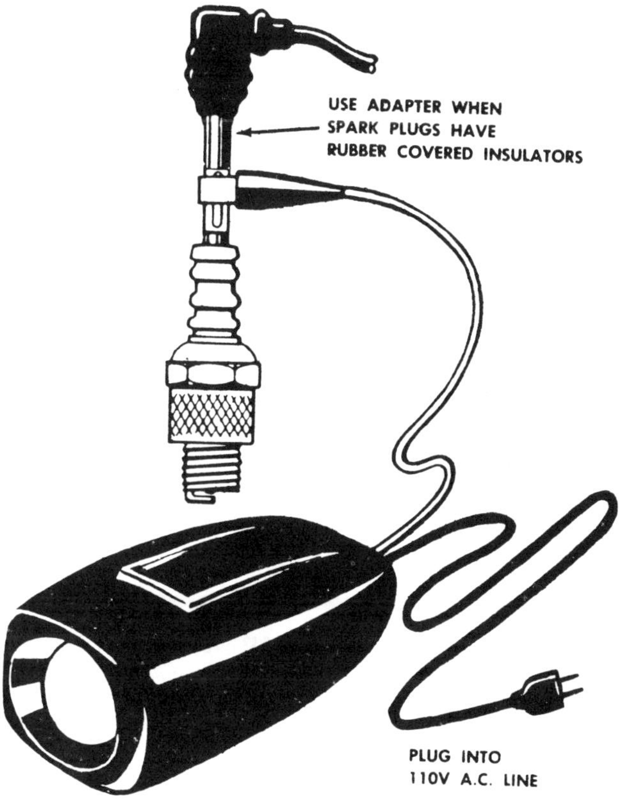

Fig. 5-4. Hook-up of AC power timing light.

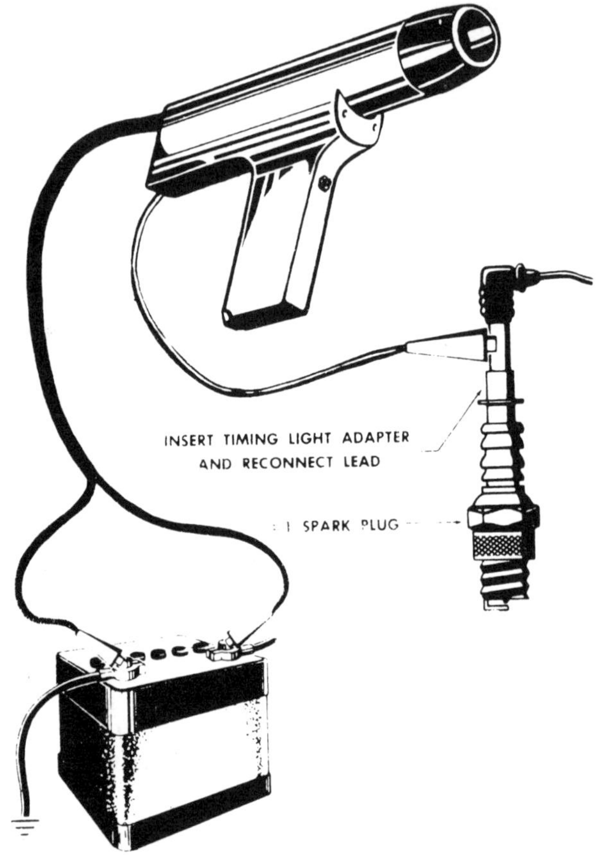

Fig. 5-5. Hook-up of DC power timing light.

TOTAL NUMBER OF CYLINDERS IN ENGINE	COMPANION CYLINDER IS:
4	3rd cylinder in firing order
6	4th cylinder in firing order
8	5th cylinder in firing order

Fig. 5-6. Companion cylinder to No. 1 for various cylinder engines.

5. Start and idle the engine. The light will flash each time the No. 1 spark plug fires.

6. Operate the engine at the specified speed and aim the timing light at the timing mark, see Fig. 5-10.

7. The position of the timing mark must agree with the manufacturer's specifications. The abbreviations TDC, ATDC, and BTDC given in the specifications mean Top Dead Center, After Top Dead Center, and Before Top Dead Center respectively. Reset ignition timing, if the timing mark is not within specification. Note that if point dwell is also going to be adjusted, it should be done prior to the resetting of ignition timing. Reconnect distributor vacuum line.

The timing mark should appear steady as the light flashes. If the mark appears to fan out or wander as the light flashes, trouble is indicated. This condition can be caused by pitted breaker points, misaligned points, improper point

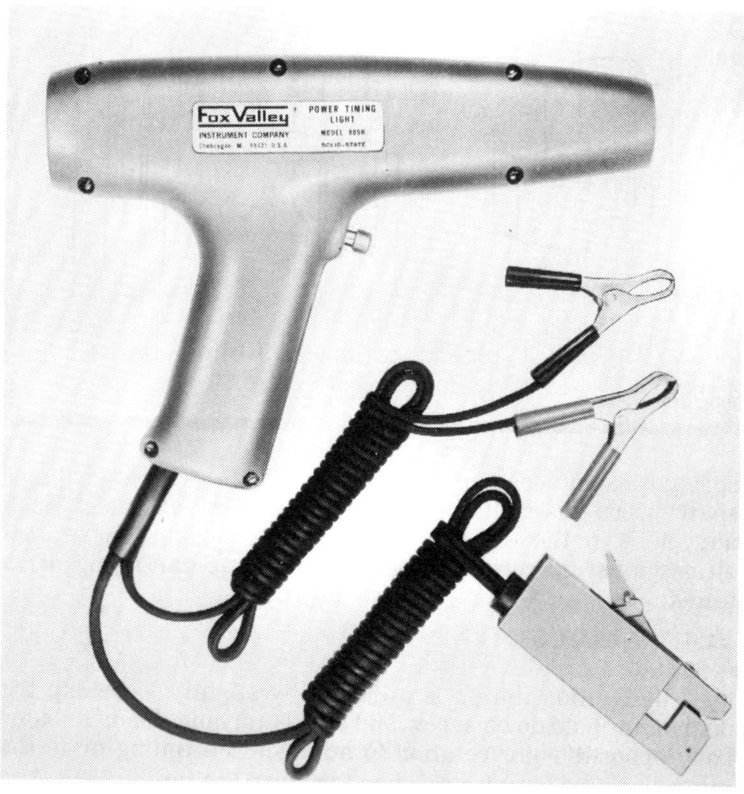

Fig. 5-7. Timing light will clamp-on pickup.

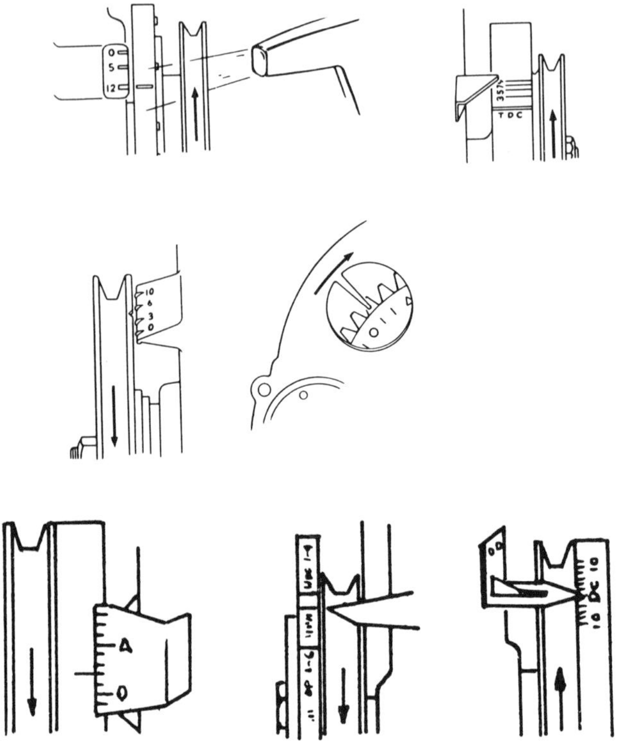

Fig. 5-8. Typical locations of timing marks.

spring tension, loose or worn breaker plate, worn distributor shaft, worn distributor shaft bushings, or excessive lash anywhere in the distributor drive mechanism. These conditions must be corrected before an engine can be properly timed.

TIMING ADJUSTMENTS

The ignition timing is adjusted by slightly loosening the distributor hold down screw and slowly turning the distributor body opposite rotor rotation to advance the timing or in the same direction as rotor rotation to retard the timing (Fig. 5-11 and 5-12). When the specified mark is aligned with the

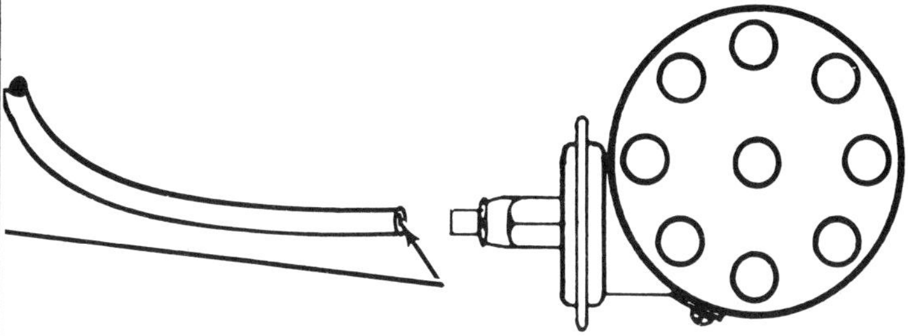

Disconnect and plug vacuum advance line. Use a piece of paper or tape to plug the end of the vacuum advance line.

Fig. 5-9. Disconnection of distributor vacuum line.

pointer, securely tighten the distributor hold down screw. Then recheck the alignment of the timing marks.

Advance Mechanism Checks

In addition to making and breaking the primary circuit, an important function of the distributor is to advance the

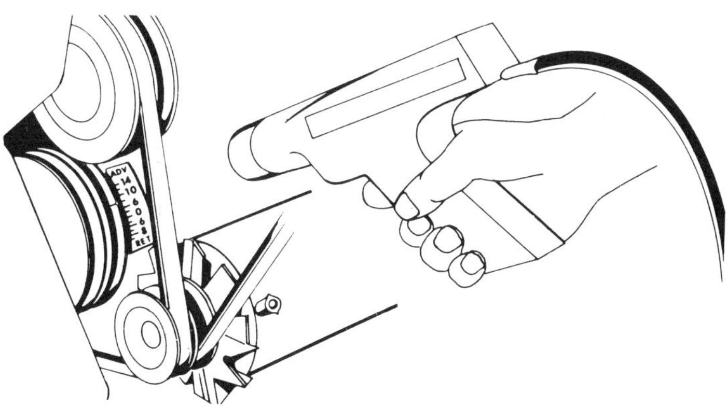

Fig. 5-10. Timing light aimed at timing mark.

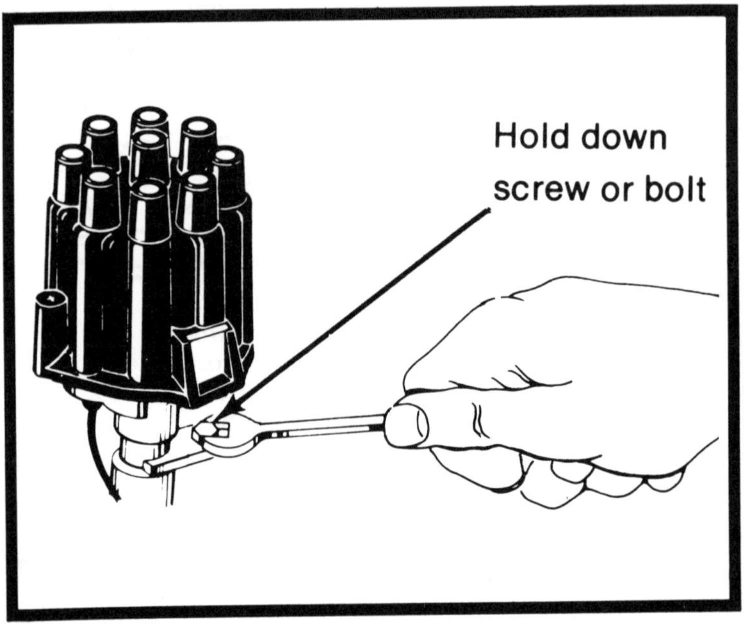

Fig. 5-11. Location of distributor holddown screw for timing adjustment.

spark timing in accordance with engine requirements as to speed and load in order to send a high tension current impulse to each spark plug at the right instant. Speed and load conditions vary considerably on the road—from light load at low speed, through heavy load at low speed, rapid acceleration, decelerating from high speed, running at uniform high speed, climbing hills, to idle.

To meet these conditions, the vacuum advance compensates for throttle opening (which corresponds to engine load) and the centrifugal advance compensates for speed. A more thorough explanation of the advance mechanisms is given in the next chapter. Both the centrifugal advance and the vacuum advance can be checked to some extent with an ordinary timing light. The exact number of degrees the ignition system advances with an increase in speed cannot be obtained by this simple timing light check. To obtain an accurate measurement of the number of degrees of ignition advance one must use an ignition advance tester, such as the type discussed in the next chapter.

The procedure for testing the advance mechanisms with an ordinary timing light is as follows:

A. Centrifugal Advance Check:

1. With vacuum advance line disconnected and engine at idle rpm, note position of timing marks.

2. Increase engine speed gradually.

3. The timing mark should move opposite to direction of rotation.

B. Vacuum Advance Check:

1. Reconnect vacuum advance line to distributor and set engine speed to about 800 rpm.

2. Suddenly thrust the throttle open and let it spring close.

3. Timing mark should momentarily move in the direction of rotation to retard spark, see Fig. 5-13.

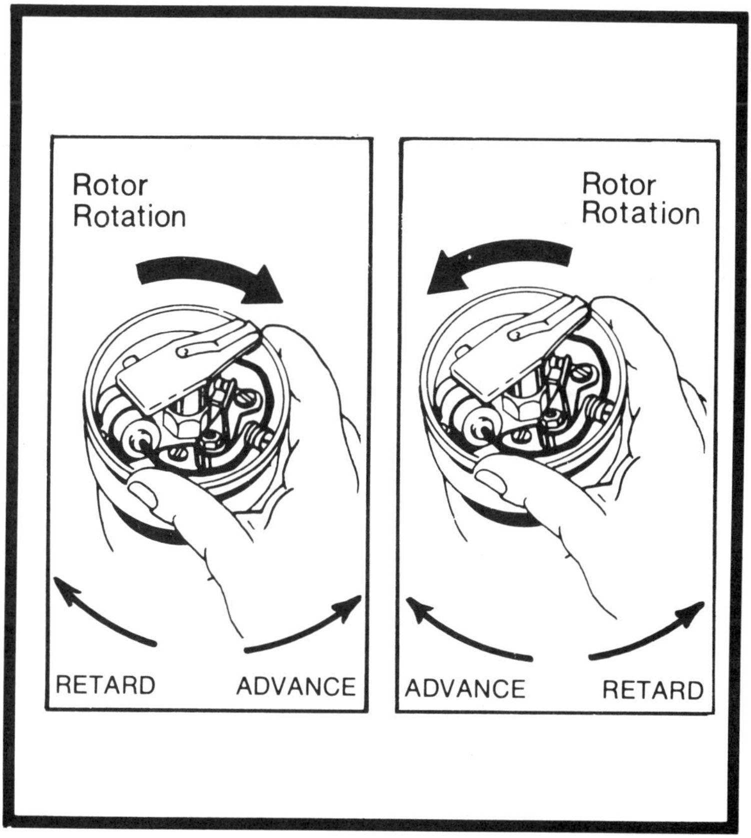

Fig. 5-12. Direction of rotation of distributor cap for timing adjustment.

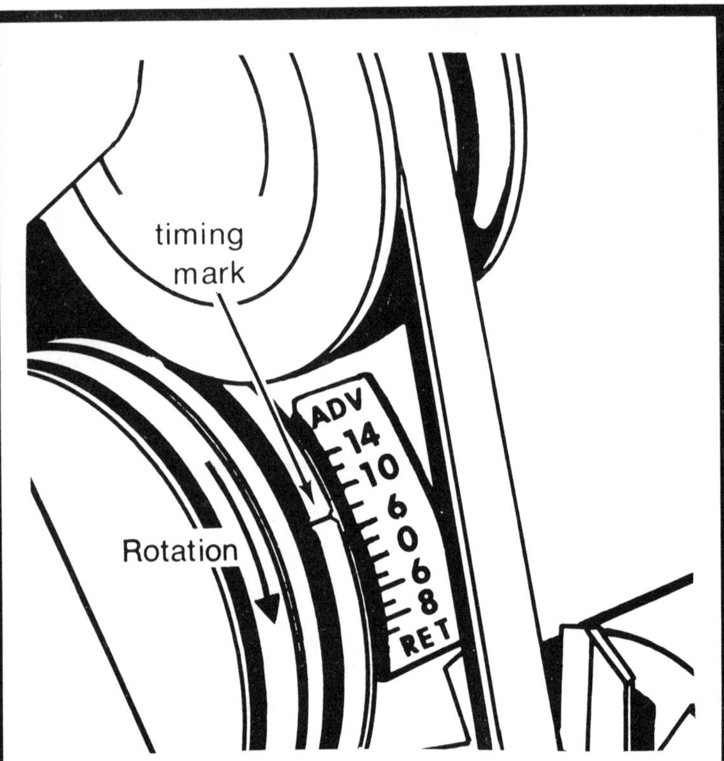

Fig. 5-13. Vacuum advance check.

Under normal speed increase, the vacuum advance control advances the spark and adds to the centrifugal advance, but a sudden increase in speed causes the vacuum to drop to a low value and momentarily retard the spark.

Chapter 6

Ignition Advance Testers

To understand the need and use of an ignition advance tester, we may begin by considering the following question. What happens to the air-fuel mixture inside the combustion chamber when a spark plug fires? The answer is that combustion of the mixture begins near the spark plug and then spreads out through the rest of the chamber. The air-fuel mixture does not explode instantaneously, but burns from one side of the cylinder to the other during a period of time.

If the mixture exploded, instead of burned, best engine performance would be achieved by having the explosion occur when the piston is at the top of its stroke. However, since it takes time for total combustion to occur, igniting the air-fuel mixture when the piston reaches the top of its stroke results in a substantial downward movement of the piston before the gases burn to the point of being able to exert maximum power against the piston. The result is a loss of power and poor fuel economy. The other extreme, igniting the air-fuel mixture too soon, results in engine overheating and possible permanent damage due to pinging and knocking. This is why the initial timing of an engine is very important. Since combustion occurs over a finite period of time, there is an optimum time before the piston reaches top dead center for ignition to occur in order to develop maximum power.

ADVANCE MECHANISMS

At high engine speeds, combustion must begin still earlier in order to have time for adequate burning. The reason for this is that the piston has increased its rate of movement, but the time required to burn the mixture has changed very little. If the piston is to be in a position where maximum power can be exerted against it, the combustion must start earlier. The function of the centrifugal advance mechanism is to fire the spark plugs at this earlier time.

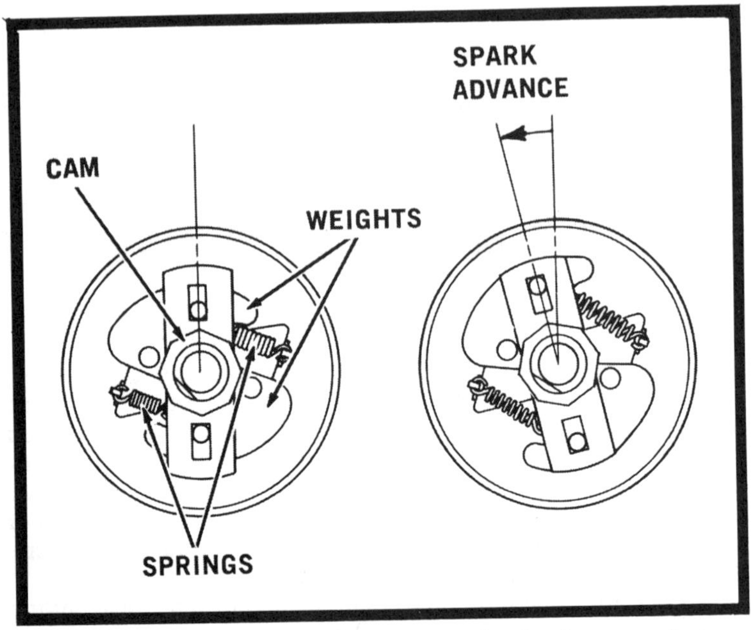

Fig. 6-1. Centrifugal advance **weights** in no-advance position (left) and full-advance position (right).

Centrifugal Advance Mechanism

The centrifugal advance mechanism, shown in Fig. 6-1, consists of weights mounted within the distributor. As distributor speed increases, the weights move away from the center of the distributor due to centrifugal force. The movement of the weights cause a rotation of the distributor cam which in turn fires the plugs earlier. When speed is decreased, springs attached to the weights return them to their initial position.

Vacuum Advance Mechanism

Why is the vacuum advance mechanism needed in addition to centrifugal advance mechanism? The answer is that the centrifugal advance compensates for engine speed, while the vacuum advance compensates for engine load. During steady part-throttle operation, less fuel is drawn into the combustion chamber than during acceleration. This smaller amount of fuel is compressed less and requires a longer time to burn. The vacuum advance mechanism, shown in Fig. 6-2, supplies additional spark advance to give this mixture more time to burn.

The changes in intake manifold vacuum control the vacuum advance mechanism. The mechanism consists of a chamber in which a flexible diaphragm is located. One side of the diaphragm is connected through a vacuum line to the intake manifold or carburetor venturi vacuum. The other side is exposed to open air. A link extending from this side rotates either the entire distributor or, on some systems, a breaker plate assembly upon which the points are mounted. With either method, the firing of the spark plugs is advanced.

The total amount of advance varies from engine to engine, but it is generally between 15 to 40 engine degrees at 2000 rpm. Separate degree readings for vacuum and centrifugal advance may be obtained from the manufacturer's specifications. These specifications are given in either engine degrees or distributor degrees. To determine the distributor degrees from a given engine degree specification, we must divide by two. For example, if the engine advance specification is 40 degrees, the distributor advance specification is 20 degrees.

If the degree scales mounted on engines and used for setting the initial timing were large enough, the amount of

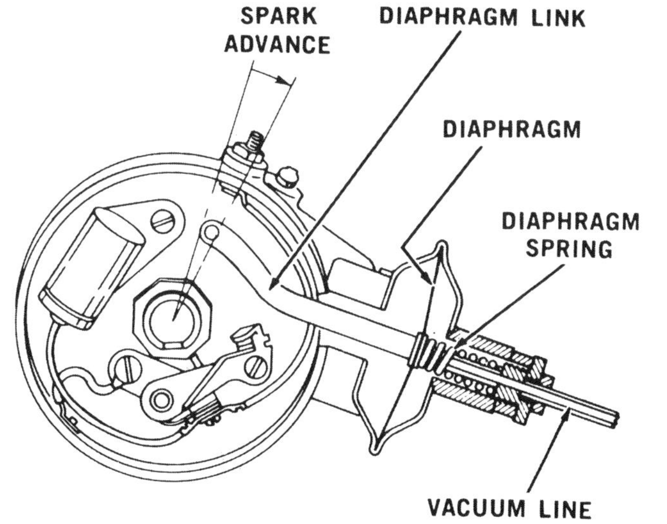

Fig. 6-2. Vacuum advance mechanism.

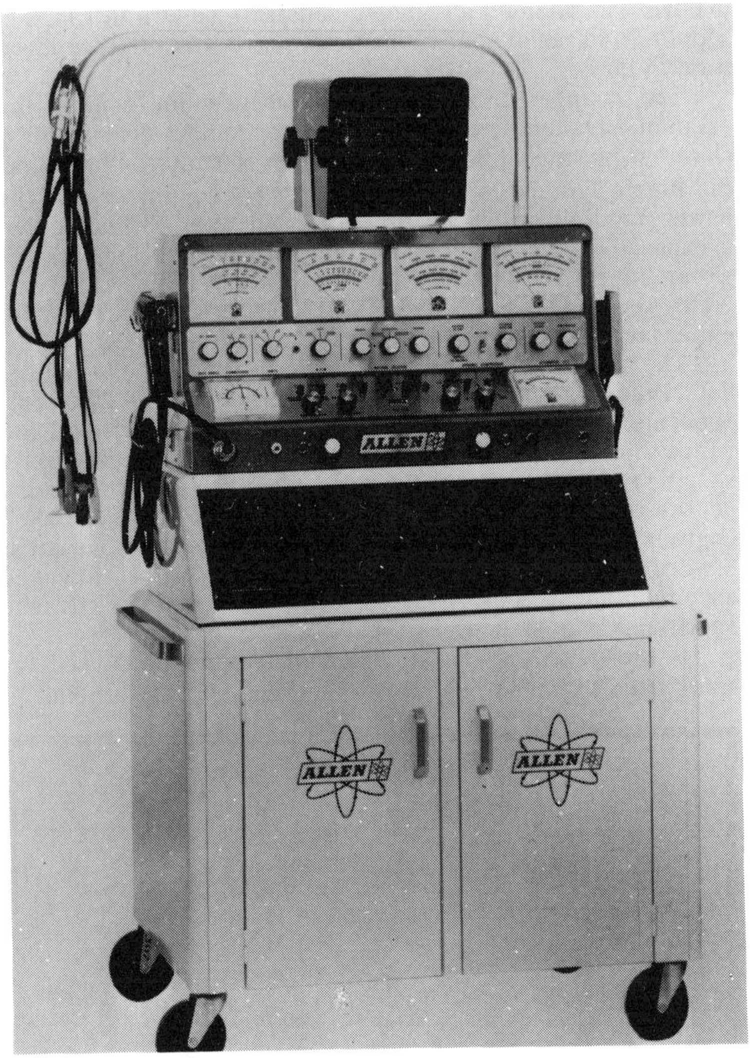

Fig. 6-3. Advance tester with advance control built into light.

advance in degrees could be read directly with a standard timing light. These scales, however, ordinarily do not go beyond 15 degrees; and, consequently, they cannot be used for making advance measurements. The ignition advance tester to be described in the remainder of this chapter is the instrument most commonly used by mechanics in testing advance mechanisms.

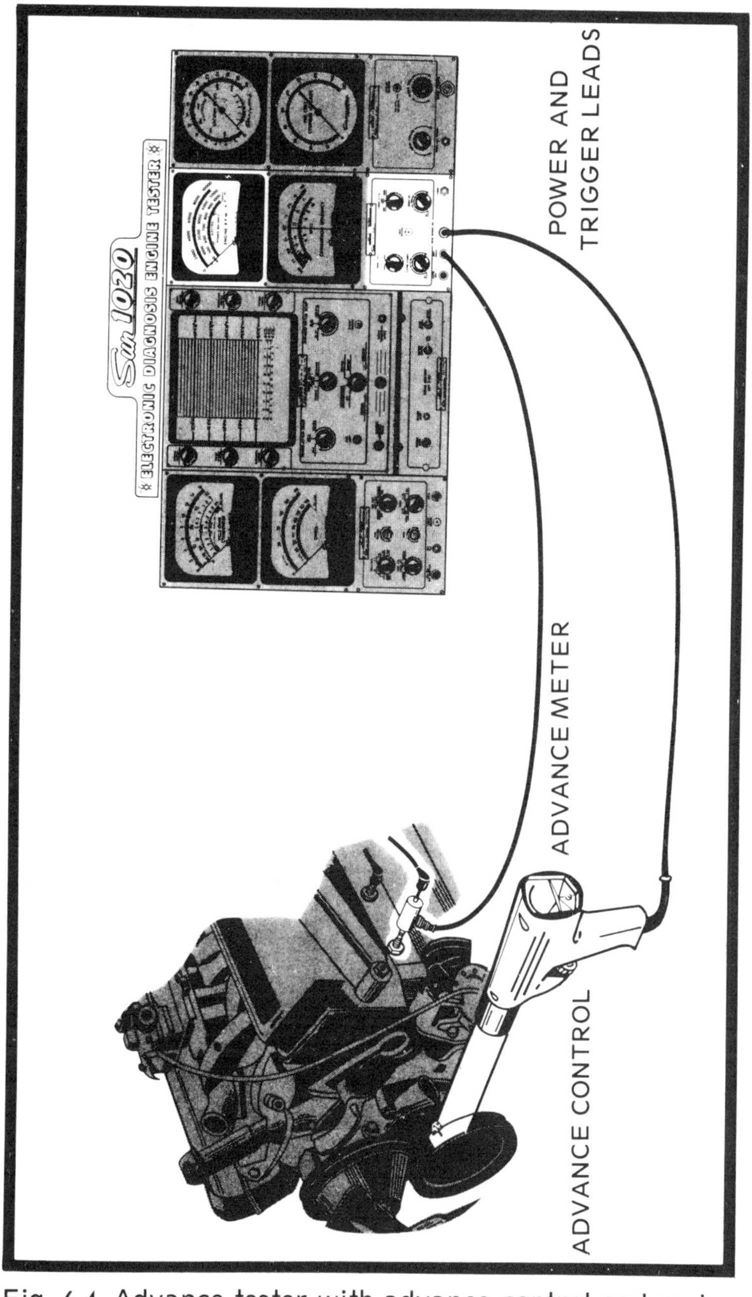

Fig. 6-4. Advance tester with advance control and meter built into light.

TYPES OF IGNITION ADVANCE TESTERS

Many manufacturers build the ignition advance tester into a console, as shown in Fig. 6-3. In this unit the advance control is built into the timing light. The console unit shown in Fig. 6-4 not only has the advance control in the timing light, but also the meter. This increases the speed and convenience with which the test can be made.

Fig. 6-5 shows a portable ignition advance tester with the advance control and meter built into the timing light. Power leads from the unit are connected to the battery in the car under test; consequently, this unit can be used anywhere in the field. The meter also has a tachometer scale on which speed may be read when the control is in the TIMING position.

The ignition advance tester illustrated in Fig. 6-6 incorporates a clamp-on magnetic pickup which lets the operator hook up the tester without stopping the engine or disconnecting any spark plug cables. In addition, instead of the meter, it has a calibrated advance control which retains the spark advance reading after the engine is stopped. The advance control is calibrated automatically for all usable speeds.

All ignition advance testers have a TIMING and ADVANCE position. When the tester is set to the Timing position it operates as an ordinary timing light similar to those already discussed in the previous chapter.

Scales

Some ignition advance testers have two scales, such as the one illustrated in Fig. 6-7. One scale is in engine degrees, the other is in distributor degrees. The two scales are related by a factor of two: if the meter reads 50 degrees of engine advance it also reads 25 degrees of distributor advance. Two scales are put on the meter because automobile manufacturer's specifications are sometimes given in engine degrees and sometimes in distributor degrees. Since it is so easy to convert from one to the other, many instrument manufacturers prefer to use only one scale on their instruments. This scale is usually engine degrees.

HOOK-UP

The hook-up of the ignition advance tester is fairly simple. First, the power leads for the tester are connected to line voltage or to the car battery, as required. Then the trigger lead is connected to number one spark plug lead. This may be by means of a clamp-on magnetic pickup or a trigger transformer. The tester is now ready for use. This is the only hookup that is needed for making both the initial timing test and the advance test.

The initial timing test is made with the selector in the Timing position. Since the procedure for setting the initial timing was described in the previous chapter, no further mention of it will be made here. We shall go directly into the advance test procedure.

ADVANCE TEST

The advance test is performed at 2000 rpm. At this speed the engine ignition is advanced and the timing mark will appear beyond the range of the indicator, as shown in Fig. 6-8.

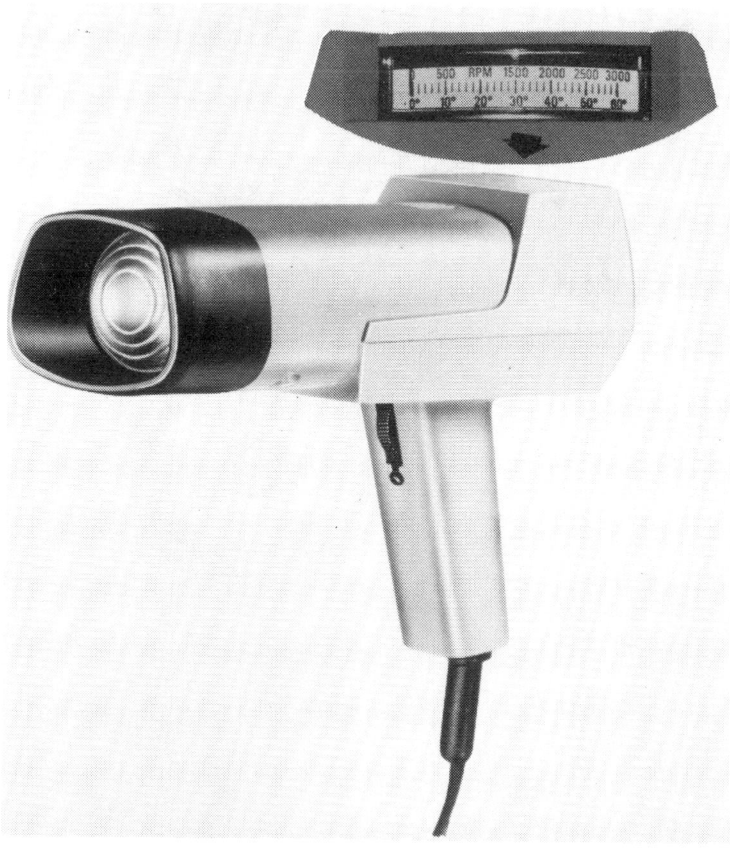

Fig. 6-5. Portable advance tester.

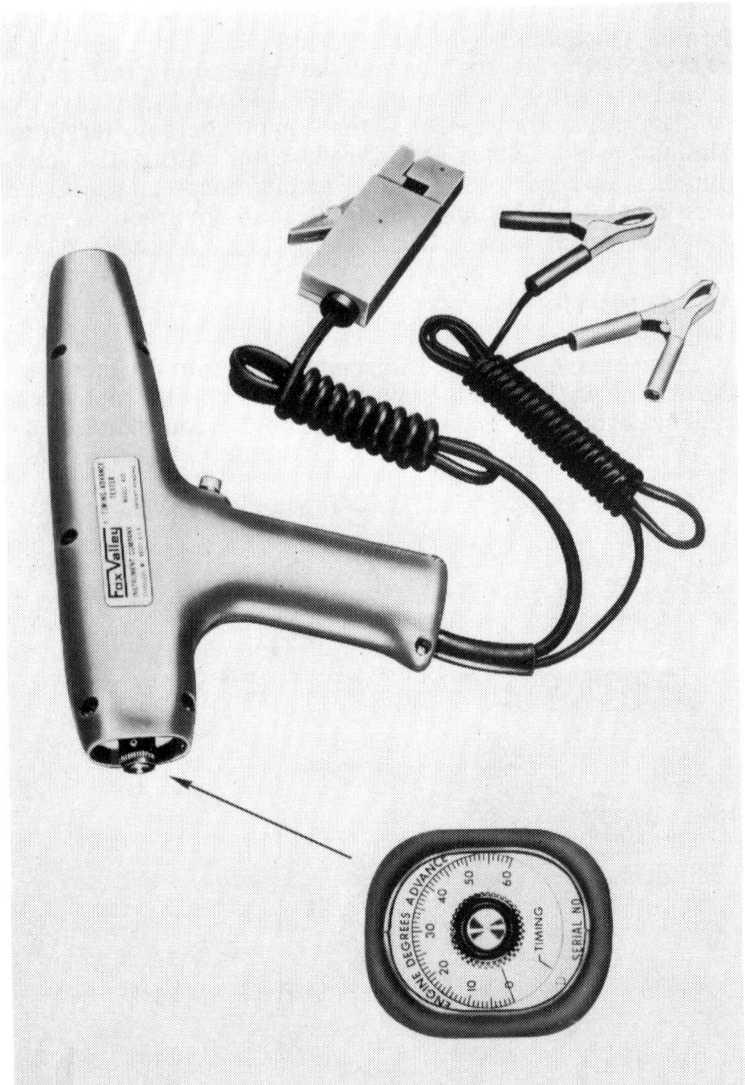

Fig. 6-6. Meterless advance tester.

The flash of the timing light is directed at the timing indicator on the engine, and the advance control on the timing light is turned until the timing mark appears in the same position as in the initial timing check. The degrees of advance is read on the ignition advance tester meter and compared with the specification.

If the advance reading is within the given specification, the centrifugal and vacuum advance mechanisms are operating satisfactorily. If the advance reading is not within the specification, a faulty centrifugal or vacuum advance mechanism is indicated. Disconnect the vacuum line and repeat the test. Then determine whether the centrifugal or vacuum mechanism is at fault by comparing the readings obtained from the test with those in the table shown in Fig. 6-9.

DISTRIBUTOR VACUUM RETARD TEST

Exhaust emission control at idle speed is often accomplished by the use of vacuum to control the action of the distributor timing retard chamber. If a vehicle is so equipped, this action may be tested by using the ignition advance tester.

First, the speed of the engine is adjusted to the rpm recommended by the manufacture for the initial timing test. Then with the ignition advance tester set for making a timing check, the position of the timing mark in relation to the timing indicator is observed. Next, the distributor vacuum advance unit retard hose is disconnected and plugged, see Fig. 6-10. With the retard hose removed, the timing mark will move to an advance position.

By rotating the advance control knob, the timing mark can be returned to the original position. With the timing mark back in its original position, the meter is read and compared with the manufacturer's specifications for engine timing

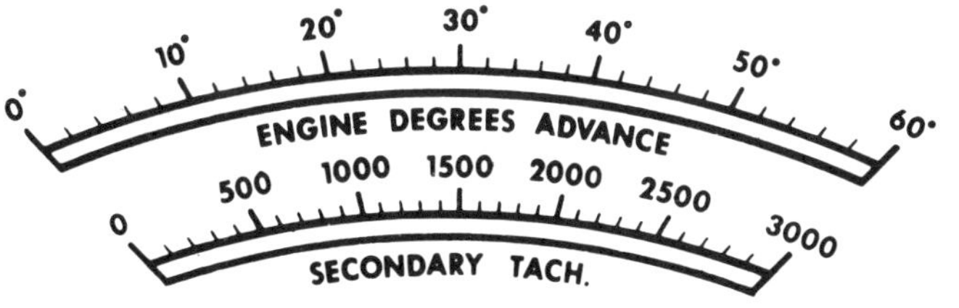

Fig. 6-7. Ignition advance tester scales.

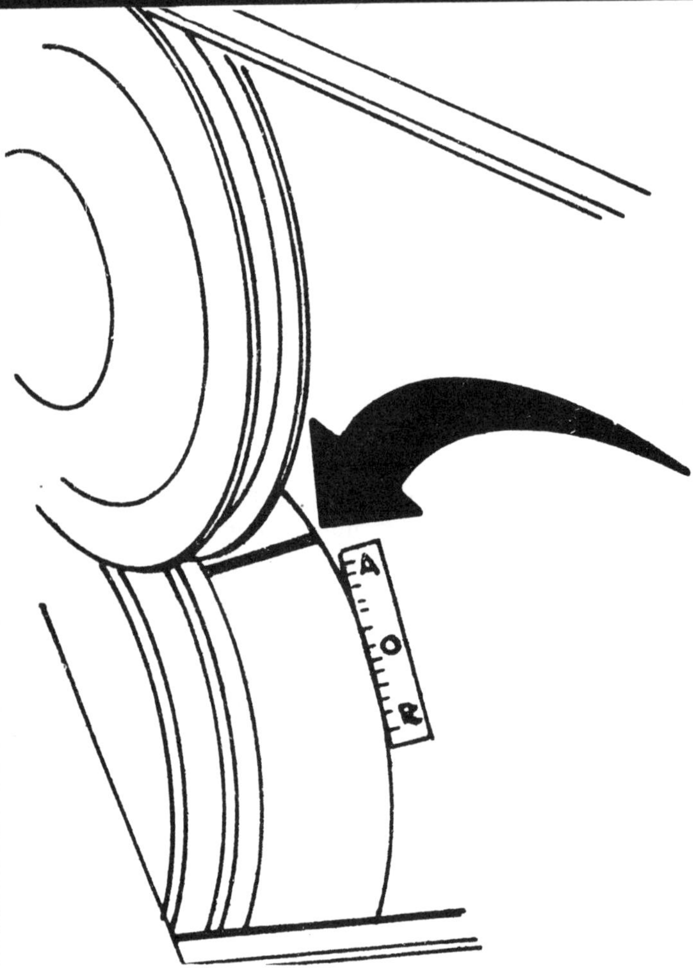

Fig. 6-8. Location of timing mark at 2000 rpm.

retard. If engine degrees of retard is not within specification, the following should be checked:

1. Plugged or leaking vacuum hose to retard port.
2. Faulty vacuum advance mechanism.
3. Sticking breaker point advance plate.

If engine timing does not change, when the retard hose is removed check for:

1. Intake manifold vacuum to retard port hose plugged or leaking.

2. Vacuum advance unit retard diaphragm leaking.
3. Malfunction of thermal vacuum valve.

HOW THE IGNITION ADVANCE TESTER WORKS

Basically, an ignition advance tester is a delay unit. It delays the flash of a timing light by an amount of time selected by the operator. As speed is increased, the timing mark on the flywheel or fan pulley pulls away from the pointer when viewed with a standard timing light, see Fig. 6-11.

The reason for this is that the firing of the spark plugs has advanced in time due to the mechanical and vacuum advance mechanisms. The ignition advance tester, when properly set, introduces a time delay in the flashing of the timing light which is exactly equal to the advance in time of the spark plug firing. The amount of time delay is set with the advance control on the unit, and is read on a meter in engine degrees.

The operation of the ignition advance tester is shown in the block diagram, Fig. 6-12. When the plug to which the trigger lead is connected fires, an impulse is sent back to the tester

METER INDICATIONS		
Vacuum Line ON	Vacuum Line OFF	PROBLEM
Meets specification	Approximately ½ of specified engine degrees advance	None. Centrifugal and vacuum advance are GOOD.
Approximately ½ of specified engine degrees advance	Approximately ½ of specified engine degrees advance	Vacuum advance is BAD
Approximately ½ of specified engine degrees advance	Below ½ of specified engine degrees advance	Centrifugal Advance is BAD
Very low advance reading	Very low advance reading	Centrifugal and vacuum advance are BAD

Fig. 6-9. Meter indications showing defective centrifugal or vacuum mechanisms.

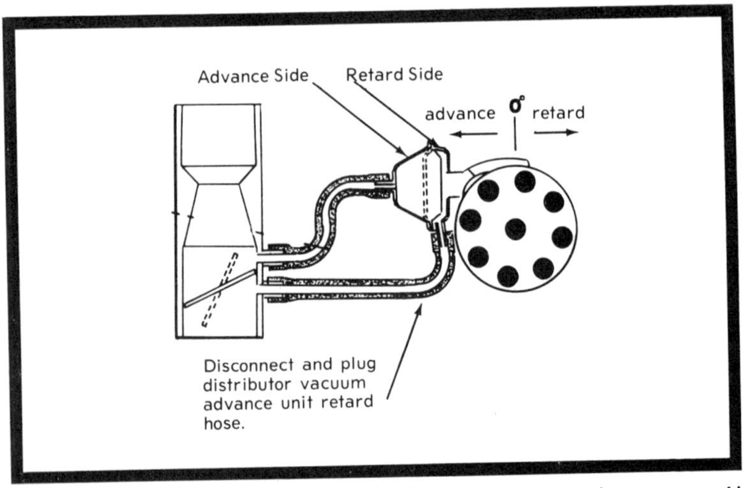

Fig. 6-10. Location of distributor vacuum advance unit retard hose.

which starts the operation of the time-delay circuit. As soon as the time delay circuit starts operating, current begins to flow through the meter. The time delay circuit continues to operate for a certain length of time, depending on the setting of the advance control, and current continues to flow through the meter until the time-delay circuit stops operating. When the time-delay circuit stops, a trigger pulse causes the timing light to flash. When the plug fires again, the sequence of events repeats itself. This sequence of events is shown in graphic form in Fig. 6-13.

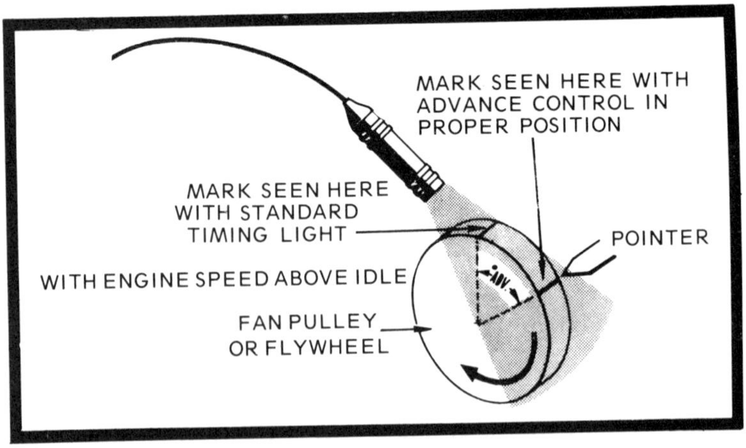

Fig. 6-11. Effect of using ignition advance tester.

Meter current flows only during the shaded areas. The dashed line represents the average amount of meter current and corresponds to some definite meter reading. The average meter current depends on the length of the TD (time delay) and the number of impulses per second. The average meter current can be increased by increasing either the TD (by rotating the advance control) or the number of impulses per second (by increasing engine speed).

Assume that an engine is operating, as shown in Fig. 6-14 at 1200 rpm. The distributor advance mechanism has caused the timing mark to be 18 degrees away from the pointer when number one plug fires. In other words, the timing is 18 degrees advanced at this speed. A conventional timing light directed at the timing mark would show it to be in this position. At this speed it will take the timing mark .0025 seconds (2.5 milliseconds) to move through the 18 degrees to the pointer. If the advance control on the ignition advance tester is so adjusted that the time delay lasts for 2.5 milliseconds also, the timing mark will appear under the pointer when the light flashes. If the time delay shown in Fig. 6-13 equals 2.5 milliseconds and the impulses per second correspond to 1200 rpm (10 impulses per second), the average meter current shown by the dashed line would cause the meter to indicate 18 engine degrees advance.

On another engine, the timing mark may be 18 engine degrees advanced at 2400 rpm, see Fig. 6-15. In this case the ignition advance tester would only have to be adjusted for a time delay of 1.25 milliseconds, since the timing mark would move through the 18 degrees twice as fast as in the previous example.

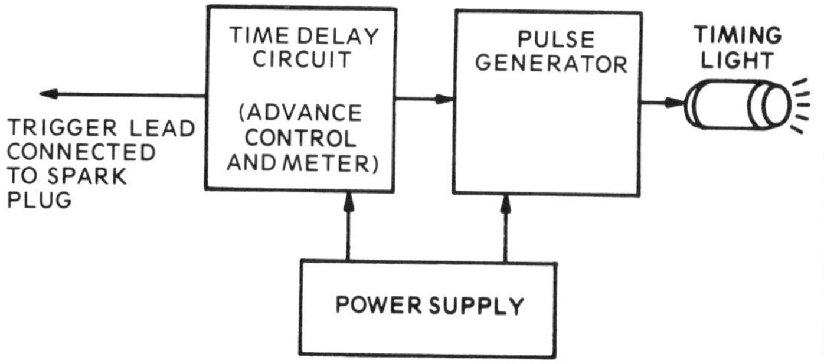

Fig. 6-12. Block diagram of ignition advance tester.

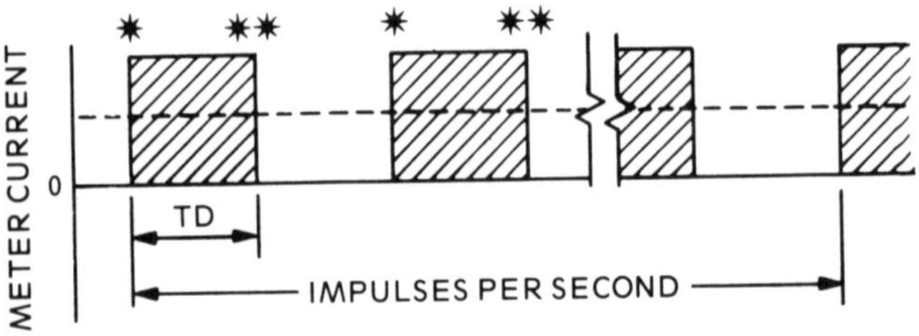

TD—Time Delay determined by setting of Advance Control
------Average meter current (determines meter reading)

* Spark occurs at plug to which pick-up is connected.
Time delay circuit starts operating.
Meter current starts to flow.

After the time delay

** Time delay circuit stops operating.
Meter current stops flowing.
Timing light flashes.

Fig. 6-13. Sequence of events in ignition advance tester.

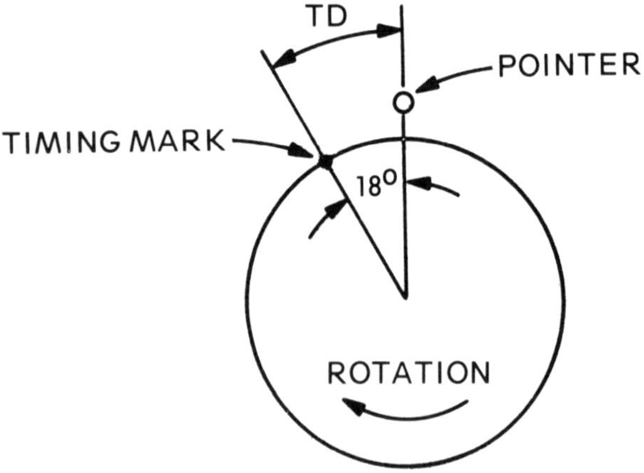

TD = 1.25 MILLISECONDS

Fig. 6-14. Operation at 1200 rpm.

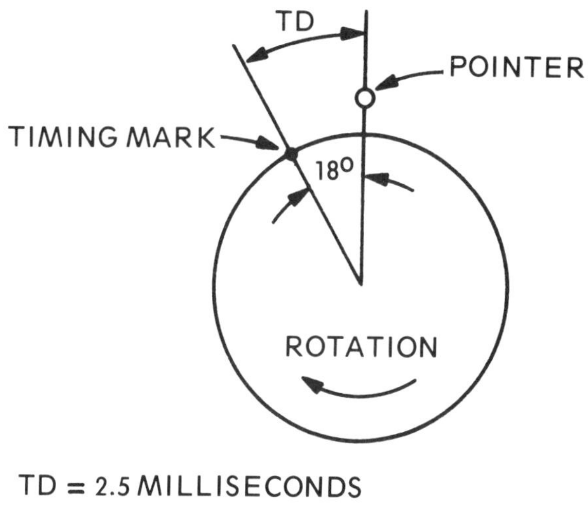

Fig. 6-15. Operation at 2400 rpm.

With the advance control adjusted for a time delay of 1.25 milliseconds, the timing mark appears under the pointer when the light flashes. Although the time delay is only half of what it was in the first example, the impulses per second are double, since engine speed is double. This means that the average meter current is the same as in the previous example; and the meter, therefore, still indicates 18 engine degrees advance, see Fig. 6-16.

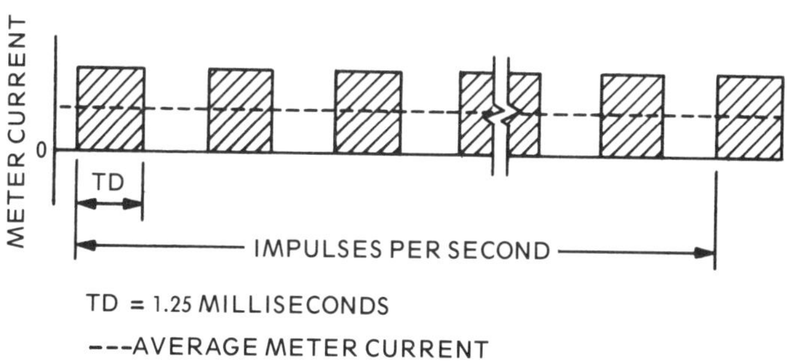

Fig. 6-16. Meter current flow at 2400 rpm.

Chapter 7

Ignition Testers

The ignition testers described in this chapter are instruments that are used for testing the coil, distributor, spark plugs, and secondary wires of an ignition system. These instruments vary in the way in which they are connected to the ignition system. One type of instrument, Fig. 7-1, connects to the primary side of the coil and permits the following ignition tests to be performed:

1. Ignition output test
2. Ignition reserve test

Another type of instrument, Fig. 7-2, connects to the secondary side of the coil, and enables the following tests to be performed:

1. Ignition output available
2. Ignition output required
3. Ignition output required under load

Still another type of instrument is connected directly to the spark plugs, where the spark current test is performed. This type is frequently combined with one of the other instruments as seen in Fig. 7-1.

In addition to describing these tests, this chapter will introduce the oscilloscope (frequently called the "scope"), since ignition system patterns must be discussed in order to understand how the ignition tester works.

IGNITION OUTPUT TEST

This test checks the ignition coil for opens, shorted turns, and internal insulation breakdown. In addition, it tests external insulation leakage of the condenser, distributor cap, rotor, and spark plug cables. To understand the manner in which this primary-connected ignition tester works, we must first describe another type of tester; the oscilloscope. Similar to a television set, the oscilloscope produces a graph-like picture on a picture tube.

Fig. 7-1. Ignition tester for connection to primary side of coil.

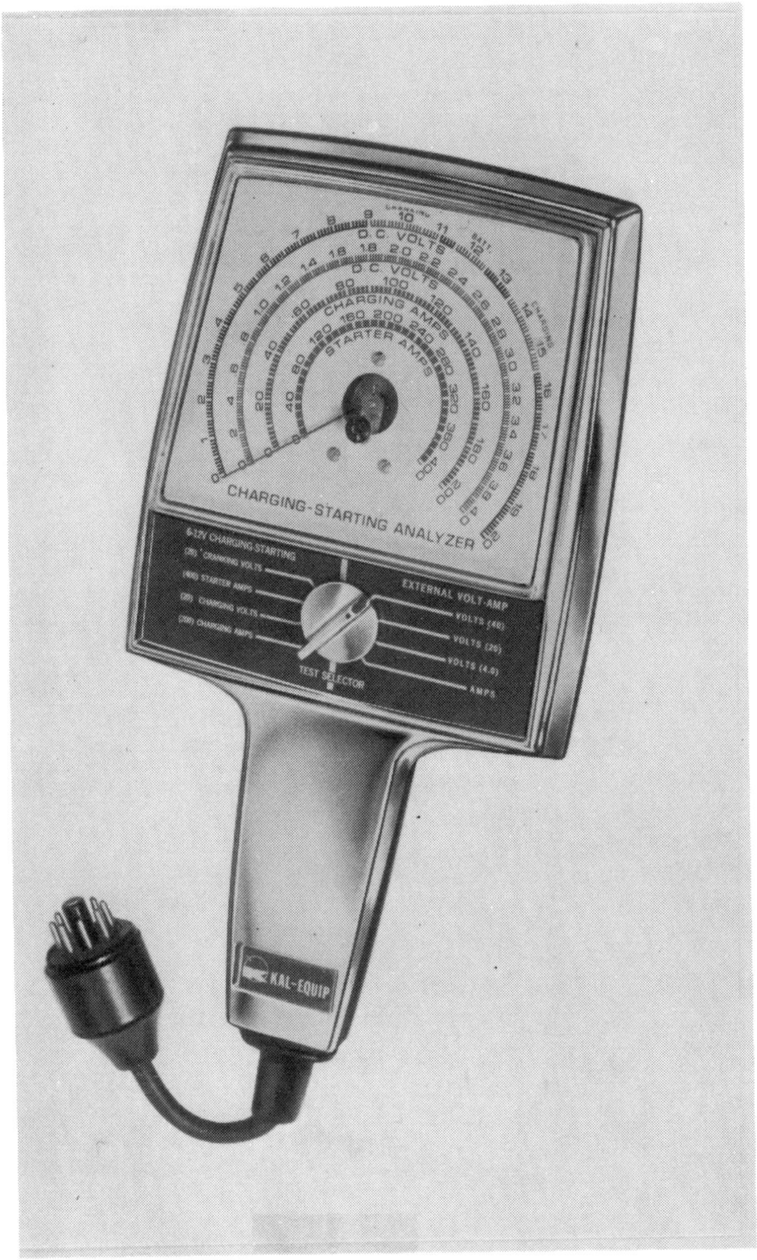

Fig. 7-2. Ignition tester for use on secondary side of coil.

The oscilloscope has long been used in laboratories and engineering departments for seeing what takes place in an electrical circuit by watching the patterns displayed on the screen of the scope. It might be said that the ignition tester is designated to measure on a meter what can be seen on an oscilloscope.

An oscilloscope is used, primarily, to show the amount, shape and frequency of the voltage it is measuring. If the test leads from an oscilloscope are connected to a 110-volt 60-Hz AC source, we will see on the screen a picture such as shown in Fig. 7-3.

The height of this curve, called a waveform or pattern, depends on the voltage at the test leads. Should the voltage being measured be increased to 220-volts AC, the curve would be twice as high. It can be seen by the shape of the waveform that the voltage rises smoothly from 0 volts to a maximum positive voltage, drops smoothly to a maximum negative voltage and then back to 0 volts again. This indicates that it is an alternating voltage. Since the voltage being measured is alternating at a frequency of 60 cycles per second, it takes 1-60th of a second to trace out this one waveform. Should the frequency of this voltage be increased to 120 cycles per second, two complete waveforms would be seen on the screen (assuming the controls of the scope haven't been changed). The closeness of the cycles on the screen would therefore indicate the frequency of the voltage being measured.

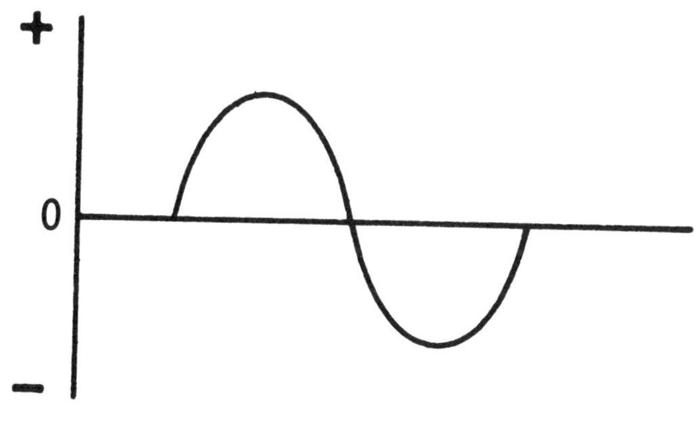

Fig. 7-3. Waveform of AC signal.

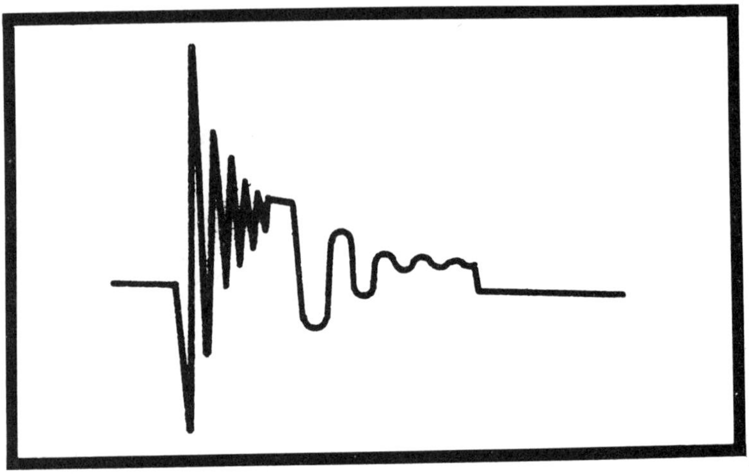

Fig. 7-4. Primary waveform for a good coil with secondary current flowing across spark plug gap.

If the test leads from the scope are connected across the negative primary terminal of an ignition coil and ground, a different type of waveform will be seen, see Fig. 7-4.

In studying this waveform, note that there are two parts to it: a high frequency portion on the left and a low frequency portion on the right. Also note that the high frequency portion is at a higher voltage than the low frequency portion. This is a normal pattern and shows that the coil is causing a spark to jump across the gap of a spark plug.

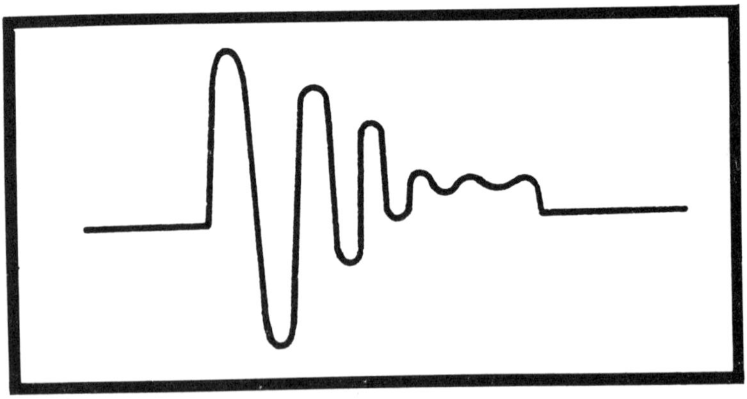

Fig. 7-5. Primary waveform for a good coil with no secondary current flowing.

If the spark plug gap is made so wide that the voltage cannot spark across it, such as when a high tension wire is completely disconnected from a spark plug, the high frequency portion will completely disappear, leaving only the low frequency portion as shown in Fig. 7-5.

It is now seen that the voltage of the low frequency portion has increased considerably and occupies the space formerly filled by both the low frequency and high frequency portion. This, also, is a normal pattern for a good coil trying to cause a spark to jump across a spark plug gap that is too wide.

Suppose that spark could jump to ground through a breakdown in the insulation of the high tension wire, even though the wire is completely disconnected from the plug. The high frequency portion would reappear and the voltage of the low frequency portion would be quite low. As seen on the scope, the waveform would again be similar to that shown in Fig. 7-4. The same situation would exist if the insulation breakdown occurred in the distributor cap, rotor, or in the coil itself.

If the oscilloscope is connected to the coil primary of a normally operating six cylinder ignition system and adjusted until six waveforms appear on the screen, each waveform will correspond to a particular plug firing. The results would be similar to those shown in Fig. 7-6.

Each of the six waveforms has its normal high and low frequency portions, showing that secondary current is flowing to all plugs. Note that the average voltage of the low frequency portion of the wave is quite low.

Pulling the high tension wire off a spark plug, number 3 for example, will change the picture from that shown in Fig. 7-6. Note in Fig. 7-7 that one of the waveforms is low frequency

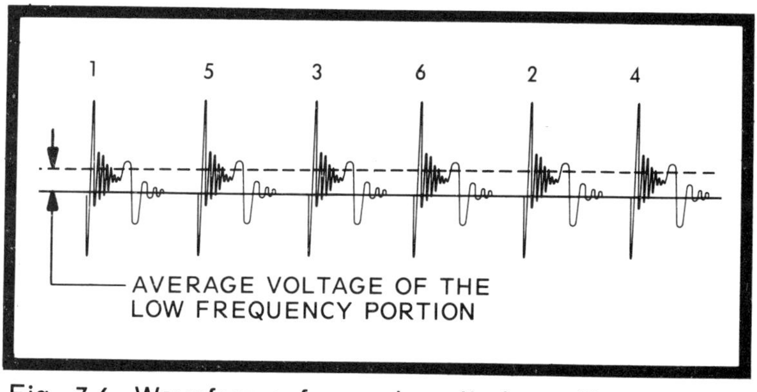

Fig. 7-6. Waveforms for each cylinder with all plugs firing.

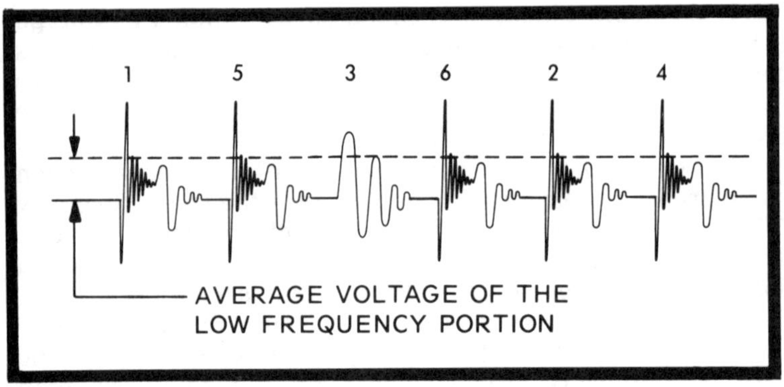

Fig. 7-7. Waveforms with number 3 plug wire disconnected.

in its entirety and that this waveform is now at a high voltage. This will raise the average low frequency voltage produced across the coil primary.

If that particular high tension wire, the distributor cap, or the rotor has been leaky, or if the ignition coil itself had been faulty, the waveforms would have remained essentially the same as shown in Fig. 7-6. In other words, the average low frequency voltage produced across the coil primary would have remained low. Only when the ignition coil and the secondary system insulation are in good condition will the average low frequency voltage rise to the value indicated in Fig. 7-7 when a high tension wire is completely disconnected from a spark plug.

Now let us look at the way in which the ignition output test operates. The circuit used to make the ignition output test consists of a tuned network and an AC voltmeter, as shown in Fig. 7-8. The test leads are connected to the distributor side of the coil and ground. The circuit is tuned only to that particular low frequency waveform that exists in an ignition primary circuit. The AC meter rectifies this low frequency voltage so that it can be measured.

The meter indicates the average voltage of these low frequency waveforms. The average voltage that exists depends upon how rapidly these waveforms are generated, which in turn depends on engine rpm and the number of distributor cam lobes. Therefore, when this test is to be performed, the engine must be operated at a specified rpm. The scale used for the test is ordinarily a GOOD-BAD band such as the one shown in Fig. 7-9.

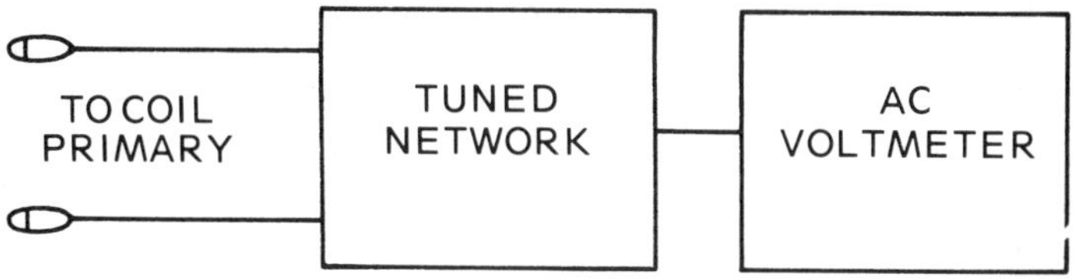

Fig. 7-8. Block diagram for ignition output test.

The test is performed by pulling a high tension wire off a spark plug. If the ignition system is in good condition, one of the waveforms will consist of only the low frequency portion and it will be at a high voltage. This will raise the average voltage resulting in a meter reading in the GOOD band.

If the high tension wire were leaky, the waveforms would have remained essentially as shown in Fig. 7-5. In other words, the average voltage of the low frequency portion would have remained low and therefore would have caused the meter to read in the BAD band.

When the first spark plug wire is removed from the engine, and the meter reads in the GOOD band, it indicates that the coil is good. Also, there is no insulation breakdown of the distributor rotor, and none in the particular segment of the distributor cap to which the disconnected spark plug wire is attached. Finally, no insulation breakdown exists in the wire removed from the spark plug.

When the No. 1 spark plug wire is replaced and each following spark plug wire is removed individually, the insulation of the removed spark plug wire is tested, and the particular portion of the cap to which that wire is attached is also tested. At the same time, a complete test is made for insulation between the rotor and any other distributor internal

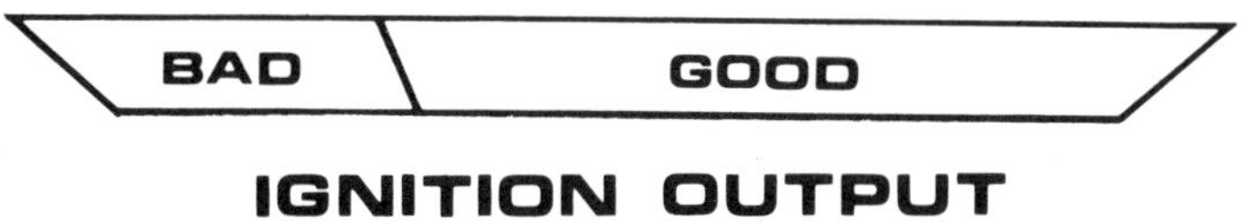

Fig. 7-9. Ignition output scale.

component (such as a condenser pig-tail lead) which might possibly be too close to the rotor.

All of these ignition test results are obtained from the ignition output test because with one spark plug wire removed, the coil voltage reaches its maximum value. This places a load upon the ignition system equal to the most severe load encountered under any type of driving conditions.

The reason the ignition output test can detect a bad coil with shorted primary or secondary turns is that such shorted turns affect the characteristics of the coil discharge, and sometimes the coil defect actually causes a spark discharge inside the coil itself. Any other coil defect which might cause the coil to absorb energy from its collapsing magnetic field would result in the coil producing less energy. This would result in a lower meter reading on the ignition output test scale.

IGNITION RESERVE TEST

The ignition reserve test tests the ability of the coil to furnish the voltage demanded by the spark plugs under the stress of light load acceleration. Light load or no-load acceleration imposes the highest ignition demands, even greater than full load operation.

When the coil cannot supply this voltage a misfire occurs. In other words, the system lacks ignition reserve. Reasons for lack of ignition reserve are: (1) insufficient coil voltage due to coil defects or defective primary circuit, or (2) excessive voltage requirements due to worn plugs or extreme resistance in plug wires.

The ignition reserve test is similar to the ignition output test already discussed in that the circuit in the tester is basically the same. The test procedure, however, is different, and the GOOD-BAD portions on the scale are reversed as shown in Fig. 7-10.

To perform the ignition reserve test, set the engine to its normal idle speed. While watching the meter, quickly and briefly accelerate the engine by rapidly pressing down the accelerator pedal and then releasing it. Do not hold the pedal down.

If the meter indication remains in the good area, no misfiring due to excessive resistance, opens, or correction exists. In addition, the spark plugs, rotor, distributor cap, and spark plug cables are all in good condition.

A meter reading in the bad area indicates a definite misfiring caused by excessive resistance, opens, corrosion, or wide spark plug gaps. To pin-point the faulty component, the spark current test, to be described later, should be performed.

IGNITION OUTPUT AVAILABLE TEST

This test is performed on the secondary side of the coil. It is a measure of the ignition circuit's maximum capabilities when an open circuit is present in the secondary circuit.

The test is performed with a peak-reading voltmeter. This is a device incorporating electronic circuitry which gives rise to a meter reading of the highest voltage in a waveform. The test leads of the instrument are connected through a voltage divider to the high voltage output of the ignition coil. With a plug wire removed from a spark plug, the maximum output the coil is capable of delivering will appear in the waveform to which the peak reading voltmeter is connected, and the meter will indicate high.

The scale used on the instrument is similar to that shown in Fig. 7-9. If the reading is in the good band, the ignition output available is satisfactory. Bad readings indicate insufficient coil output resulting from bad points, a defective coil, high resistance primary connections, or reversed coil polarity.

If there is a doubt as to the quality of the secondary circuit ignition components, each spark plug cable should be removed from the spark plug, one at a time, and held away from ground. The meter pointer should move to the good area as each cable is removed.

If a reading is low on a particular cylinder, remove the spark plug cable at the distributor cap. A leaky cable will cause the indication to rise; a distributor cap defect will cause the indication to stay low.

IGNITION OUTPUT REQUIRED TEST

This test is used primarily to detect the presence of above-normal ignition requirements. One method of performing this

Fig. 7-10. Ignition reserve scale.

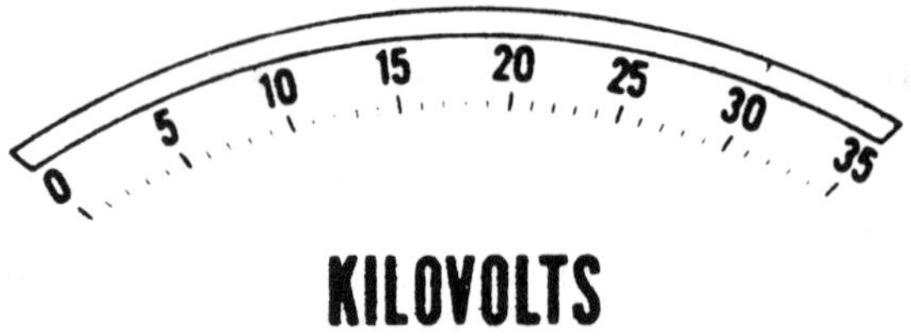

Fig. 7-11. Ignition output required scale.

test is to use both the peak reading voltmeter described under the ignition output available test and the cylinder shorting device discussed in the chapter on the electronic compression tester.

With the peak reading voltmeter connected to the output of the coil and with all the spark plug wires in place, the peak reading voltmeter will indicate the voltage of the plug that requires the maximum voltage for firing. If this voltage is higher than it should be, as measured on a kilovolt scale, such as shown in Fig. 7-11, then the cylinder shorting device can be used to locate the faulty cylinder. The cylinder shorting device can be set to short-out any selected plug in the firing order. When the cylinder demanding the high voltage is shorted, the meter indication will fall to a lower level. The causes for the high voltage can be a defective spark plug cable, distributor cap tower, or spark plug.

If the cylinder shorting device is rotated through the firing order and no fall-off is noted on the meter, then a centralized defect such as an improperly seated or defective coil-to-discriminator-cap cable is indicated. Where cables have been improperly seated, always check coil and distributor cap towers for burning or carbon tracking.

Another method of performing the ignition output required test is to use the peak-reading voltmeter to measure each individual plug. The instruments that perform the test in this manner incorporate a circuit similar to the cylinder shorting device, except that instead of using the circuit to short a plug and prevent it from firing, the circuit is used to gate the peak-reading voltmeter, so that it will only read the maximum voltage appearing in a period of time correspon-ding to the cylinder selected for the test. In this manner, all the individual cylinder voltage requirements can be obtained and compared. If one or more voltages are higher than others, trouble is indicated.

IGNITION OUTPUT REQUIRED UNDER LOAD

This test primarily detects worn-out spark plugs. It is made with the peak-reading voltmeter connected to the secondary side of the ignition system. The accelerator is suddenly depressed to the floor and quickly released while observing the meter pointer. The meter reading should go up, but a reading higher than specified indicates worn spark plugs.

SPARK CURRENT TEST (SECONDARY RESISTANCE TEST)

The spark current test determines the amount of resistance in the secondary system by measuring the amount of current which flows to each of the spark plugs. As in any electrical circuit, the amount of current which can flow is affected by the amount of resistance in the circuit. In the ignition secondary system, excessive resistance takes the form of broken wires and poor or corroded connections.

To perform the spark current test, spark plug adapters are connected to each spark plug, as shown in Fig. 7-12. Then with one lead connected to ground (this frequently is the red lead), touch the other lead to each spark adapter in turn while the engine is operating at a speed of approximately 1500 rpm.

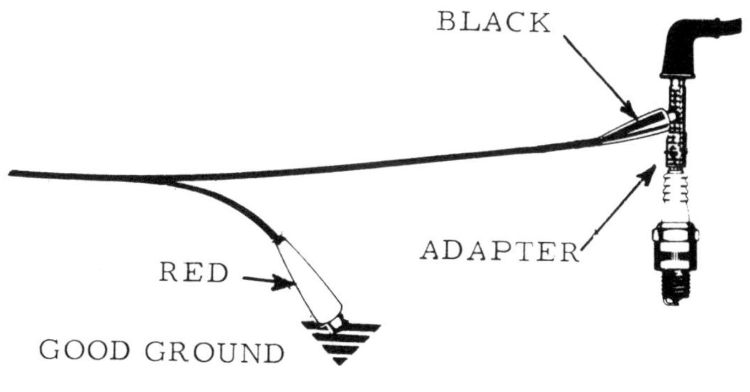

Fig. 7-12. Connection of spark plug adapters.

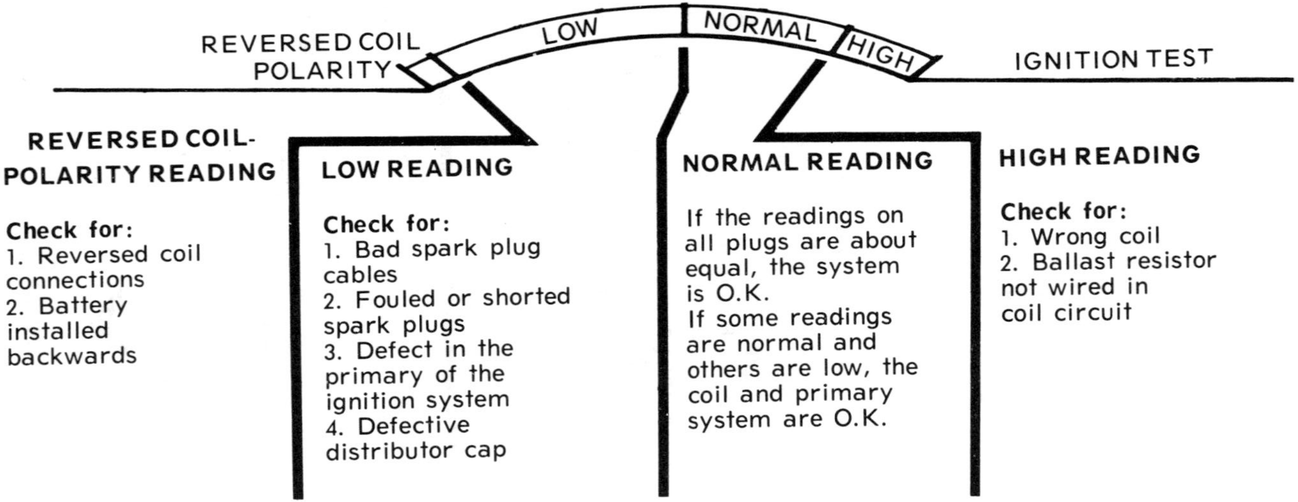

Fig. 7-13. Spark current scale and meter interpretation.

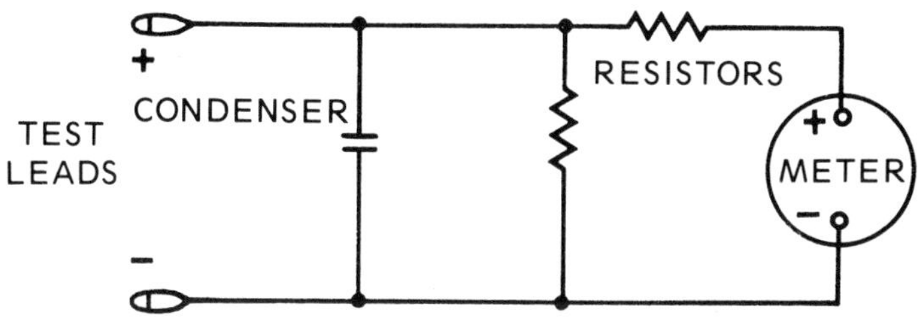

Fig. 7-14. Circuit diagram of spark current tester.

The spark plug to which the lead is connected will miss since the secondary current will now flow through the tester and register on the meter.

The spark current test is primarily a comparative test; that is, the readings should be similar at each spark plug. As a rule, no definite meter readings are specified. The meter band may be marked with areas representing acceptable and unacceptable readings, see Fig. 7-13. Meter interpretations are also illustrated in Fig. 7-13.

A coil will produce a slightly better spark and the spark plug gaps will not widen as fast when the current is negative at the spark plug terminals. If the reading on the meter is in the reverse direction, the coil polarity is reversed. To correct this condition, reverse the primary connections at the coil, but first determine whether or not the battery is installed backwards.

The circuit for one type of spark current tester is shown in Fig. 7-14. As can be seen, the circuit is somewhat similar to a voltmeter circuit. The condenser in this circuit smooths the incoming pulsating secondary current to provide a higher steady meter reading. Whenever the spark current test is performed with an ordinary voltmeter, the readings will be lower because of the absence of the capacitor.

To perform the test with a voltmeter, the red (positive) voltmeter lead is connected to ground and the black (negative) lead is touched, in turn, to each plug. Note the relative scale reading on the 0-16 volt scale. If the meter reads backwards, spark polarity is reversed (probably due to reversed connections on the coil primary). Normally the readings should be 4 or more with no more than one major division variation between spark plugs.

Chapter 8

Ignition Oscilloscopes

Scope testing is easy. Use of the scope can be quickly learned and the results can be quite satisfactory. Even a novice may uncover a very subtle problem in the ignition system, such as a defective rotor, the very first time he uses a scope. Naturally, such an experience will make a confirmed scope user from a skeptic of scope testing. More than likely, once you have used a scope yourself, it will always be the first tool you turn to when an ignition problem arises.

Scopes come in many sizes and shapes, but they all do basically the same thing. The scopes shown in Fig. 8-1 and Fig. 8-2 vary in screen size from 3 inches to 23 inches. The smaller scope uses a special electrostatic tube while the larger scope uses an ordinary magnetic tube of a type frequently found in television sets. With either scope, the same defects in an ignition system can be found.

As indicated in the previous chapter, the graph traced on the scope screen has two elements, since it is showing the relation between them as changes take place. The two elements are voltage and time. The ignition voltage goes through some interesting changes during each four-stroke cycle of each cylinder, and the time element is the period during which each cylinder fires once. With an engine running at 3600 rpm, the time is one-thirtieth of a second, which will give some idea of how fast the instrument functions.

Because the electron beam has virtually no mass, it can follow instantaneous variations in the ignition voltage; whereas a meter, having mechanical parts to move, is an averaging device.

IGNITION SCOPE CONTROLS

All ignition scopes are similar in that they all have common controls and leads. Some of the names for the controls and leads may vary from scope to scope, but their function remains the same. Fig. 8-3 shows a general front panel layout of the controls for an ordinary scope. The FOCUS and

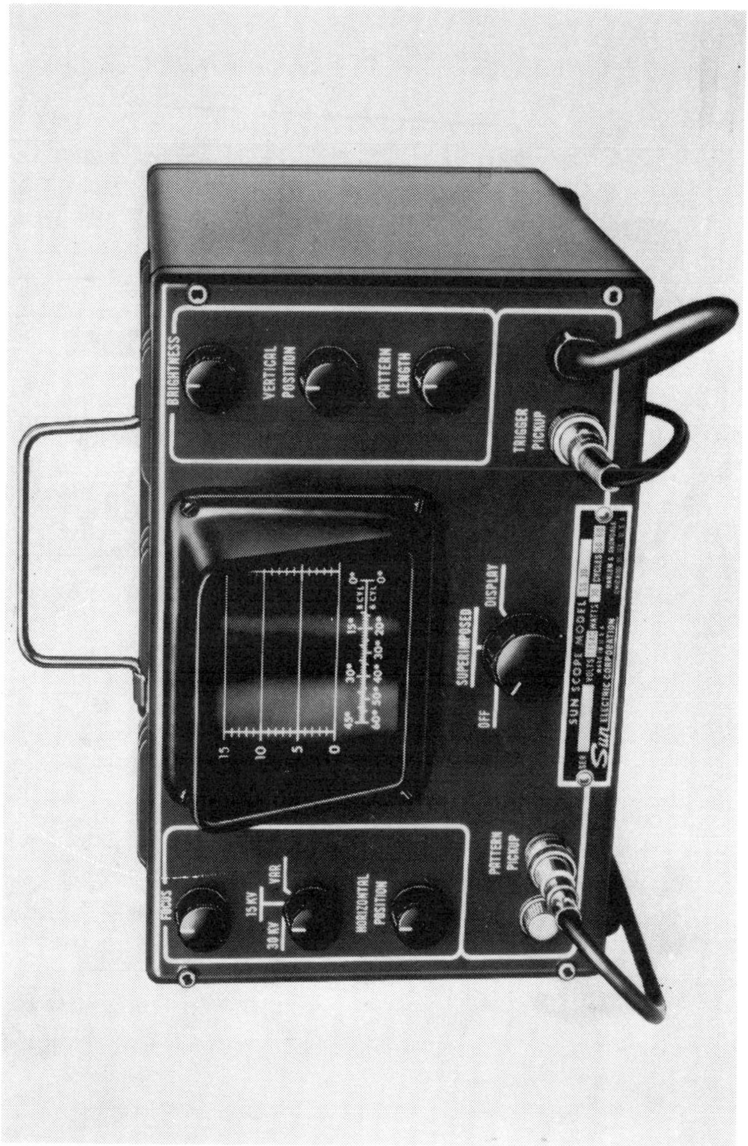

Fig. 8-1. Ignition scope with 3-inch picture tube.

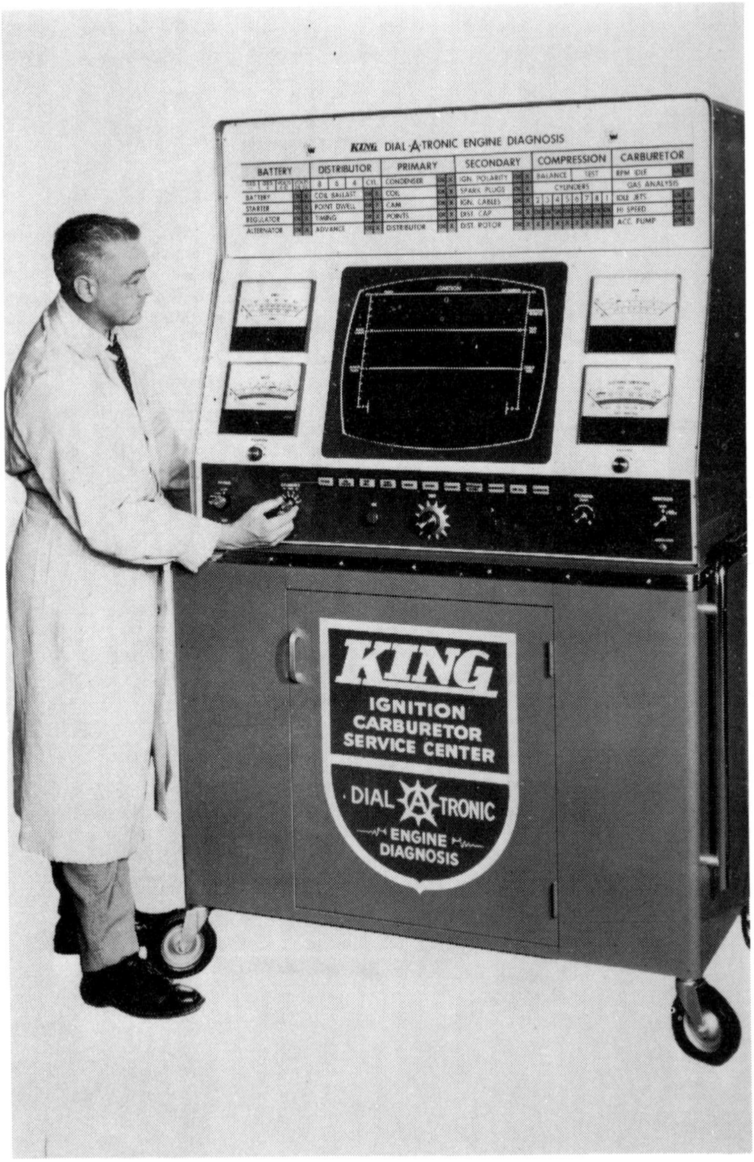

Fig. 8-2. Ignition scope with 23-inch picture tube.

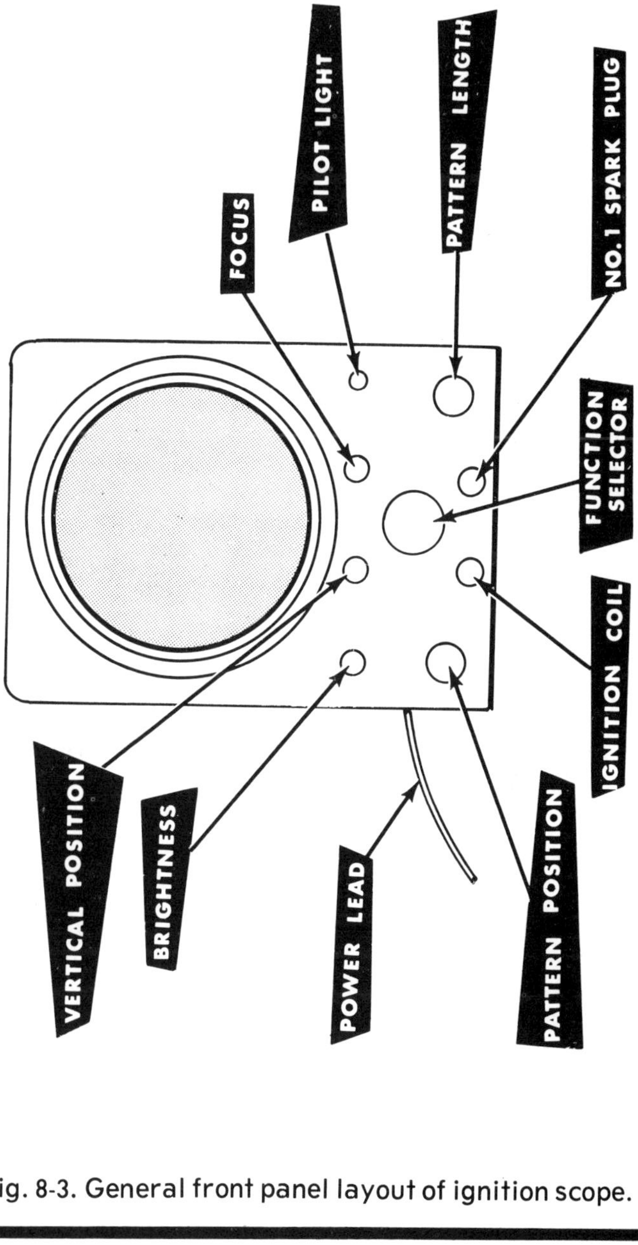

Fig. 8-3. General front panel layout of ignition scope.

BRIGHTNESS controls adjust the clearness and brilliance of the pattern respectively. The VERTICAL POSITION control adjusts the position of the pattern on the screen either up or down. The PATTERN POSITION control adjusts the position of the pattern either to left or to right. The PATTERN LENGTH control adjusts the horizontal size of the pattern. On some Scopes, although not shown in the illustration, a HEIGHT control permits variable adjustment of the vertical size of the pattern.

The POWER LEAD is connected to either line voltage or a battery depending on the type of scope, and the PILOT LIGHT indicates when the scope is on or off.

No. 1 SPARK PLUG lead (sometimes called TRIGGER PICKUP) is connected between the distributor and number one spark plug and triggers the scope so that 4, 6, or 8 cylinders appear in the scope display, depending upon the engine under test.

IGNITION COIL lead (sometimes called PATTERN PICKUP) is connected between the coil and distributor and provides the secondary signal that is displayed on the scope for analysis.

On some scopes, an additional PRIMARY lead is used to display the signal appearing on the primary side of the coil. This lead is connected to the distributor side of the coil and ground.

In Fig. 8-4 is shown a typical FUNCTION SELECTOR that allows the operator to select either the superimposed, raster, or display pattern. Above the FUNCTION SELECTOR is a CIRCUIT SELECTOR for the selection of either the primary or secondary pattern. These patterns will be explained in detail in the following sections.

BASIC SECONDARY SCOPE PATTERN

To interpret scope test results, it is important that the basic scope waveform or pattern be thoroughly understood. It is generally a good idea, when studying scope patterns, to consider them to be graphs of voltage with respect to time. The vertical displacement from the zero line (either up or down) represents voltage at any instant.

Many scopes are designed to display either a secondary pattern, which is obtained at the output of the coil high tension tower, or a primary waveform, which is obtained at the distributor primary terminal of the coil. However, because the secondary pattern is generally the most informative for

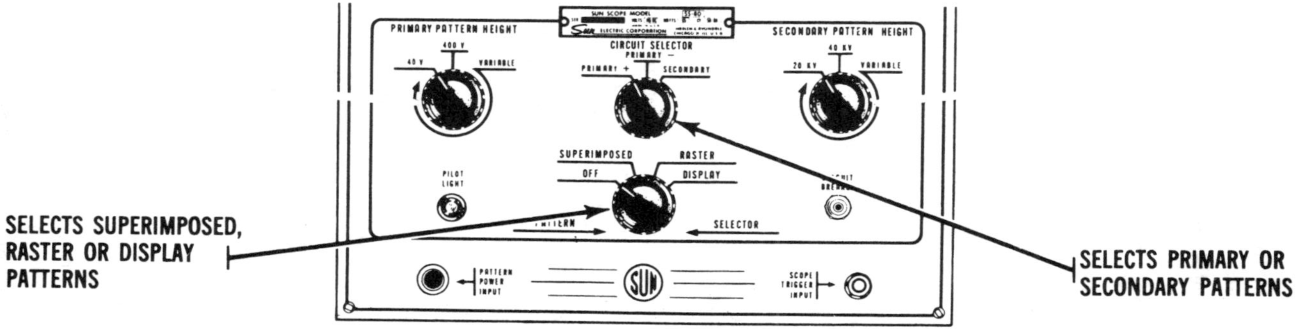

Fig. 8-4. Detail view of function selector.

showing over-all ignition system operation, this is the pattern that should be best understood.

Each part of the secondary pattern represents a specific phase of ignition system operation. For the purpose of understanding and analyzing the scope pattern, it is divided into sections as shown in Fig. 8-5 and described as follows:

Spark Voltage - (Sometimes called FIRING LINE.) Indicates the voltage required to break down the plug and rotor gaps. Spark voltages for all cylinders should be from 5 to 15 kilovolts.

Spark Line - Indicates the voltage required to maintain the spark. The spark line should be level for unsuppressed systems and tilted slightly upwards toward the spark voltage for suppressed systems (carbon wires, resistor cap, rotor, or plugs).

Coil-Condenser Action - Shows the dissipation of the remaining energy after a plug has fired. Look for several oscillations that gradually diminish in size.

Point Close - At this time, the breaker points close. Look for a downward line followed by several oscillations.

Point Open - At this time, the breaker points open, resulting in the production of high voltage. Look for an abrupt 90 degrees angle.

Dwell Angle - Represents the period of time during which the breaker points are closed. The dwell angle is read directly in degrees on a dwell scale on the screen.

Shown in Fig. 8-6 is the normal parade (sometimes called display) pattern of an eight cylinder engine. The display on the scope screen follows the firing order of the engine. Number one cylinder appears on the left side of the waveform. Following this is the display of number two cylinder in the firing order, etc. The last spark voltage seen on the right hand side corresponds to number one cylinder. The rest of the pattern for number one cylinder is seen, as mentioned above, on the left hand side.

Shown in Fig. 8-7 is the normal secondary raster pattern of a six cylinder engine. The raster pattern makes use of the vertical size of the scope screen by stacking the patterns of each cylinder, one above the other, so that all of the individual patterns are vertically distributed on the screen. This allows the scope patterns of individual cylinders to be displayed at full screen width and permits detailed, close-up inspection of all engine cylinders simultaneously. The raster pattern is especially helpful for comparing cylinders with one another and for finding variations in the patterns. The numbers next to the cylinders in Fig. 8-7 show an example of an engine firing order 1-5-3-6-2-4.

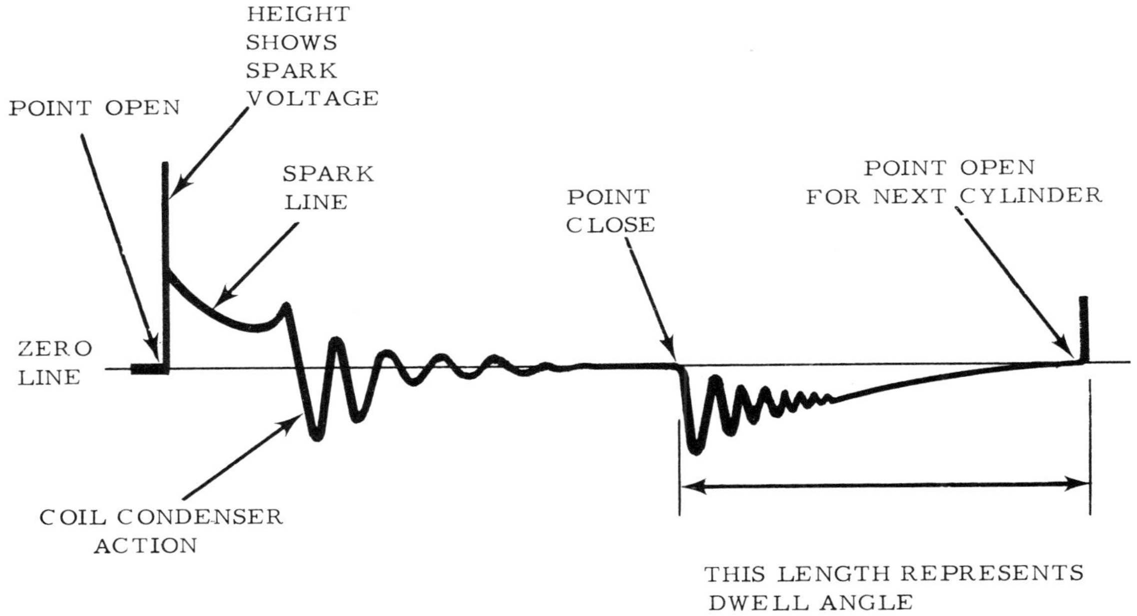

Fig. 8-5. Secondary scope pattern.

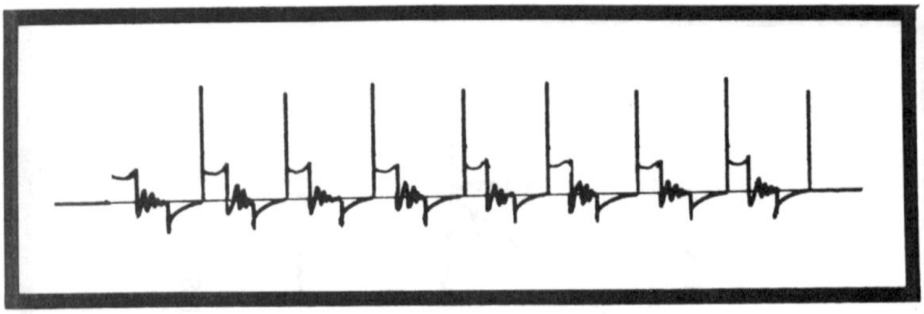

Fig. 8-6. Normal parade pattern for eight-cylinder engine.

Shown in Fig. 8-8 is the normal secondary superimposed pattern for an 8-cylinder engine. The superimposed pattern is the display obtained by simultaneously placing the patterns of all cylinders one on top of the other. Superimposing scope patterns is advantageous in many instances, because it provides a convenient method of testing the ignition system for over-all uniformity. By expanding the pattern horizontally to completely fill the screen, any variation in the basic pattern sections can quickly be detected.

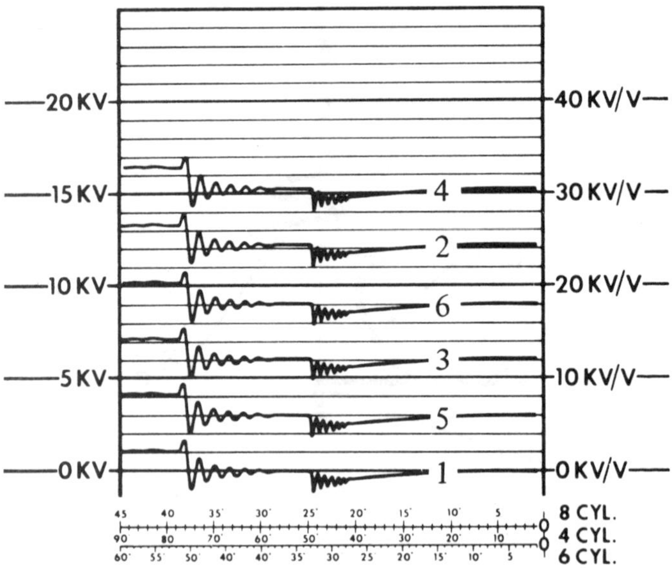

Fig. 8-7. Raster pattern of six-cylinder engine.

Since the secondary pattern is obtained from the high voltage output side of the coil, the voltage, as measured from the zero line, is in kilovolts (abbreviated KV). One kilovolt is 1000 volts; therefore, 30 kilovolts means 30,000 volts. The scope screen will have a scale on which voltages to 30 or 40 kilovolts can be measured.

BASIC PRIMARY SCOPE PATTERN

Since any voltage variation in the ignition primary circuit will be reflected in the secondary circuit because of transformer action, it is seldom necessary to use a primary pattern for general ignition testing. If desired, however, a primary pattern can be displayed on most scopes.

The KV scale on the scope screen applies only to the display of secondary patterns. Although a resemblance exists between primary and secondary patterns, it should be noted that the voltage values represented in the primary patterns are considerably lower than those represented in the secondary patterns because of the ratio of the primary to secondary windings in the ignition coil. Some scopes have a primary voltage scale on the screen of from 300 to 400 volts. The primary pattern, shown in Fig. 8-9, has sections similar to the secondary pattern:

Spark Signal - The opening of the points cause the magnetic field of the coil to collapse which creates a high secondary voltage that fires the spark plug.

Spark Plug Firing Section - This represents the period of time that current continues to flow across the spark plug gap.

Spark Extinguish Signal - Current flowing across the spark plug gap ceases to flow at this time.

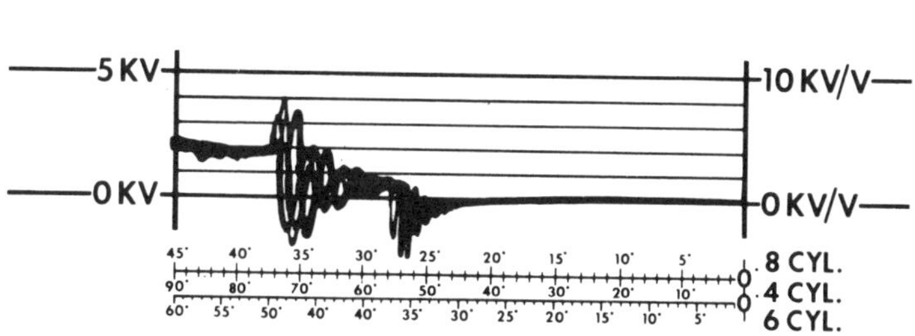

Fig. 8-8. Secondary superimposed pattern.

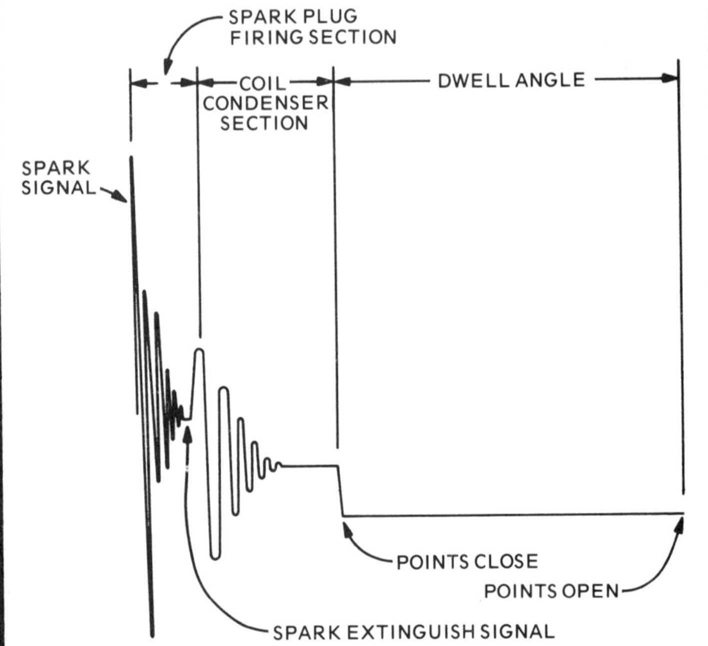

Fig. 8-9. Primary scope pattern.

Coil-Condenser Section - Unused spark energy is dissipated in the coil and condenser after the plug stops firing.

Point Close - At this time the breaker points close. One should see an abrupt 90 degree angle.

Dwell Angle - Span of time that points remain closed.

Point Open - At this time the points break contact and the cycle repeats.

The primary parade, superimposed, and raster patterns are shown in Figs. 8-10, 8-11 and 8-12.

TRACE BRIGHTENING

The spark voltage in the secondary pattern has already been discussed; and, as will be seen later, this is a very important part of the pattern in scope testing. The spark voltage consists of a very short duration pulse which ordinarily produces a very faint image that is difficult to see on the scope screen. This is because the electron beam sweeps across the phosphor on the scope tube so quickly that it does not illuminate fully. To overcome this problem special trace brightening circuits have been developed.

One method of trace brightening is to introduce a signal of the proper polarity into either the grid or cathode of the oscilloscope tube. This modulates the intensity of the beam during the short pulse duration of the spark voltage, thereby providing a brighter presentation of the pulse. The result of this method is a flaring of the tip of the spark voltage, as illustrated in Fig. 8-13.

Another method involves the formation of an entirely new pattern, Fig. 8-14. This is done by rapidly charging a capacitor during the short duration of the spark voltage pulse, and then slowly discharging the capacitor to zero before the next spark voltage pulse occurs. The start of the slow discharge is at the maximum height of the spark voltage. This new pattern reveals only the variations in height of the spark voltage and therefore makes it easy to compare spark voltages with one another.

HOOK-UP

The hook-up of all scopes is similar, Fig. 8-15. A power lead is connected to a battery or line voltage, depending upon whether the scope is DC or AC operated. A trigger lead is connected to number one spark plug. A pattern lead is con-

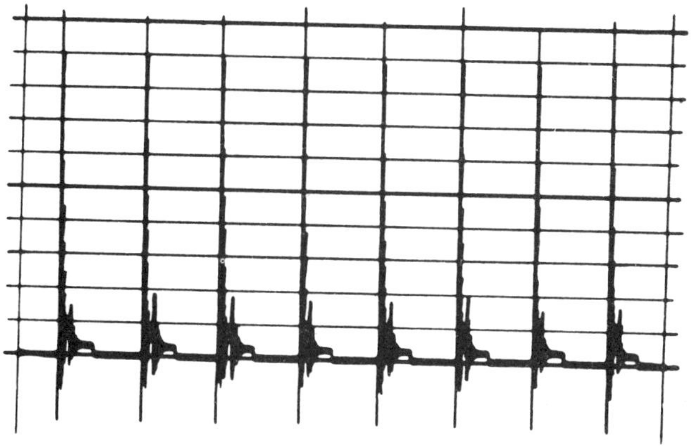

Fig. 8-10. Primary parade pattern.

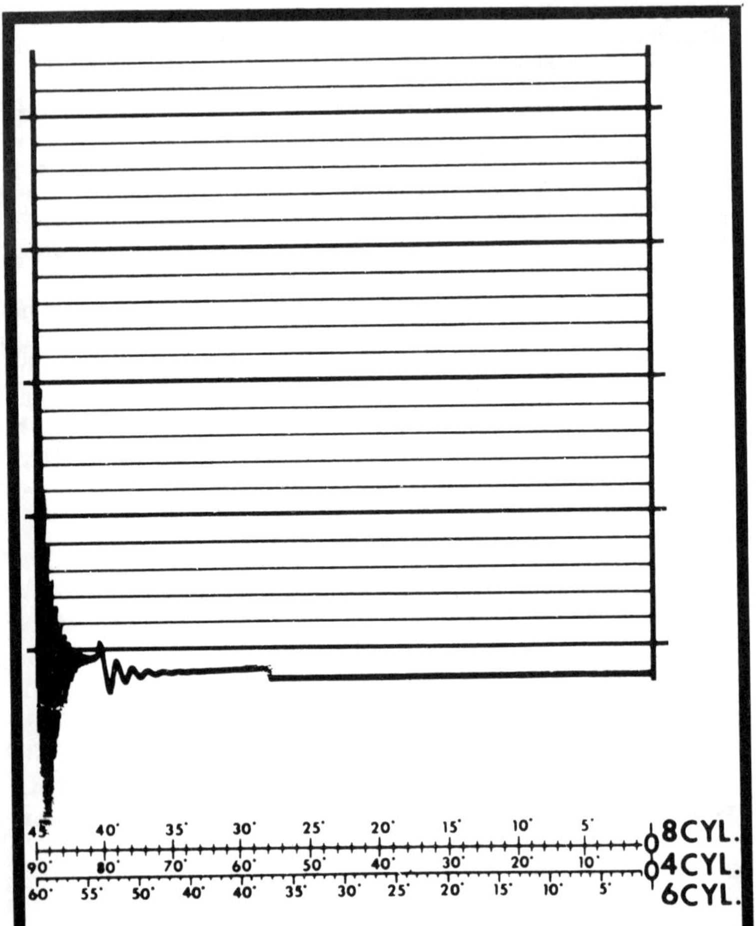

Fig. 8-11. Primary superimposed pattern.

nected to the output of the coil tower. A ground wire is connected to the engine block. Finally, on scopes that display a primary pattern, a primary lead is connected to the distributor side of the coil.

TEST SEQUENCE

The advantage of scope testing is that once a mechanic is familiar with the basic scope patterns and the patterns that indicate defects, he can instantaneously spot a defect in the

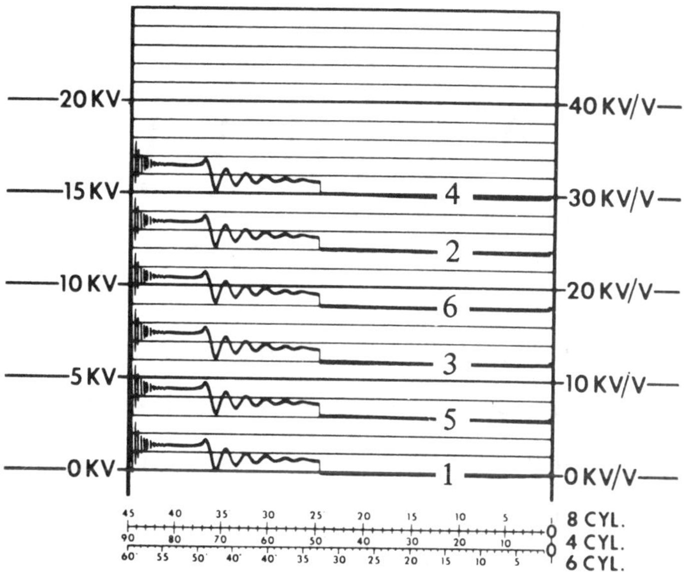

Fig. 8-12. Primary raster pattern.

ignition system by a glance at the scope. This, however, is speaking of a person who is experienced in scope testing: the beginner needs some guides or procedures to follow to feel certain about the thoroughness of his work. For this reason, the following sequence was developed.

This sequence examines in detailed steps the various portions of the scope pattern and shows the affect on the

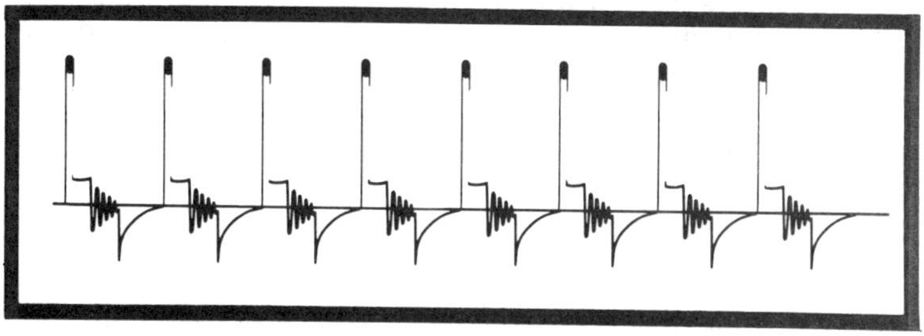

Fig. 8-13. Trace brightening by increasing intensity of spark voltage.

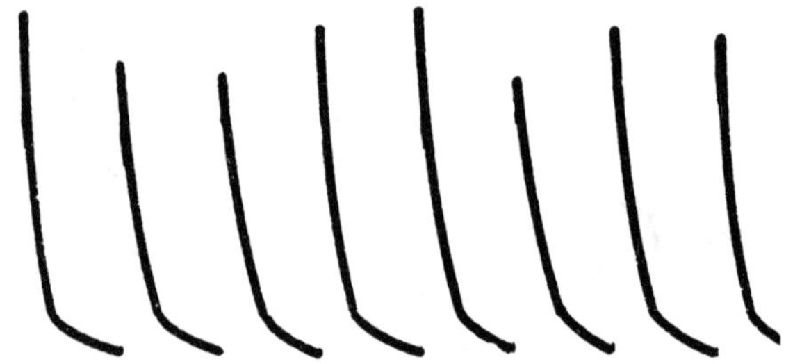

Fig. 8-14. Trace brightening by means of new pattern.

pattern of defects in the ignition system. The tests are as follows:

1. Spark polarity test.
2. Available voltage test.
3. Spark insulation test.
4. Spark voltage test.
5. Spark voltage under load test.
6. Dwell test.
7. Coil-condenser test.
8. Point action test.
9. Secondary resistance test.
10. Variation-in-dwell test.

Some of the steps can be performed with both the secondary and primary patterns and some of them can be performed with the raster pattern. The objective here is not to show all of the ways in which a particular test can be made, but to explain fundamentals from which one can learn. For this reason, the simple secondary pattern is used to illustrate all of the steps in the procedure. Once the fundamentals are understood, any variation in testing procedures or any unique feature on a particular scope will be easily mastered.

Spark Polarity Test

Incorrect spark polarity can result in up to 40 percent more voltage being required to fire the spark plugs. Consequently, misfiring of the spark plugs may occur, particularly at higher speeds. The normal and the unsatisfactory patterns for the spark polarity test are shown in Fig. 8-16. The normal pattern is in the upright position, while the unsatisfactory pattern is inverted. If the pattern is unsatisfactory, one should check for:

1. Coil connections reversed
2. Incorrect coil
3. Battery connections reversed

Fig. 8-15. Hook-up of ignition scope.

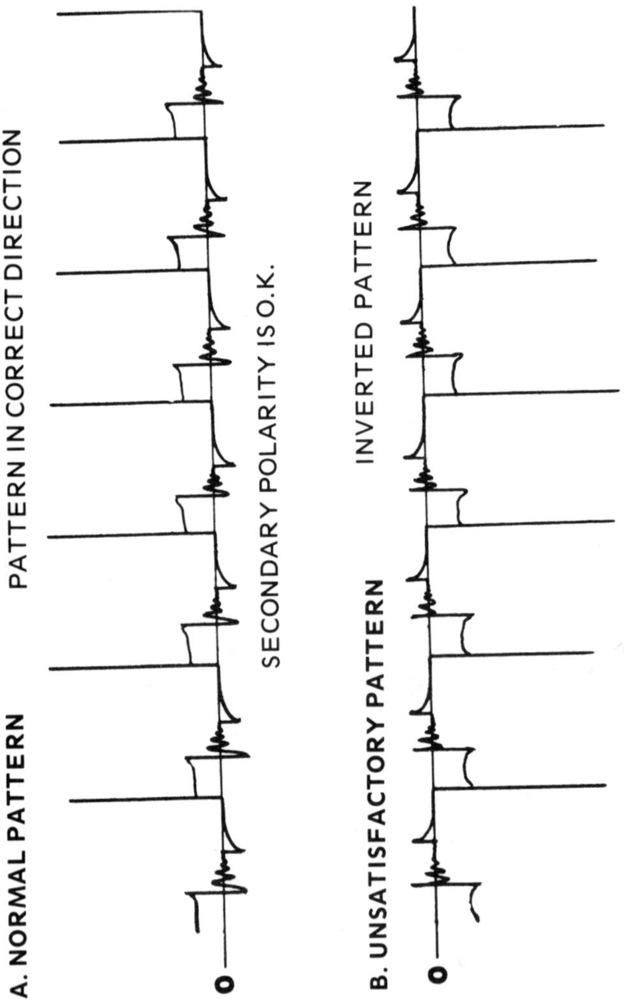

Fig. 8-16. Scope patterns for spark polarity test.

The normal pattern is the same for both positive and negative ground systems.

Available Voltage Test

When a spark plug wire is removed from a spark plug, the secondary voltage rises to the highest value the system is capable of producing. Comparing this value to the firing voltage (the voltage required to overcome the rotor and spark plug gaps and to establish a spark across the spark plug electrodes) gives a good indication of ignition reserve. If ignition reserve is low, ignition misfiring may occur whenever the vehicle is operated under severe conditions.

To perform the test, use insulated pliers and remove the wire from any spark plug except the one to which the trigger lead is connected. The wire may be removed at either the distributor end or spark plug end. The voltage of the open-

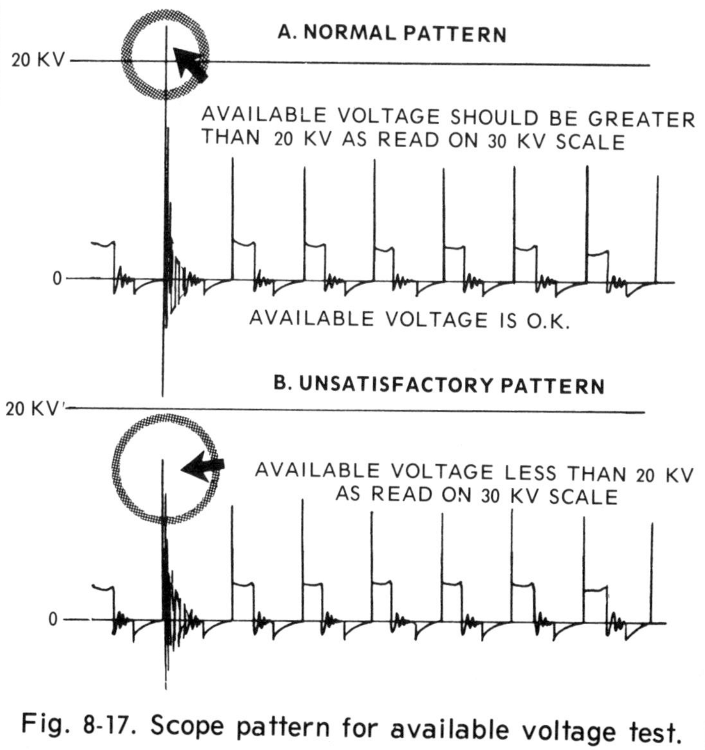

Fig. 8-17. Scope pattern for available voltage test.

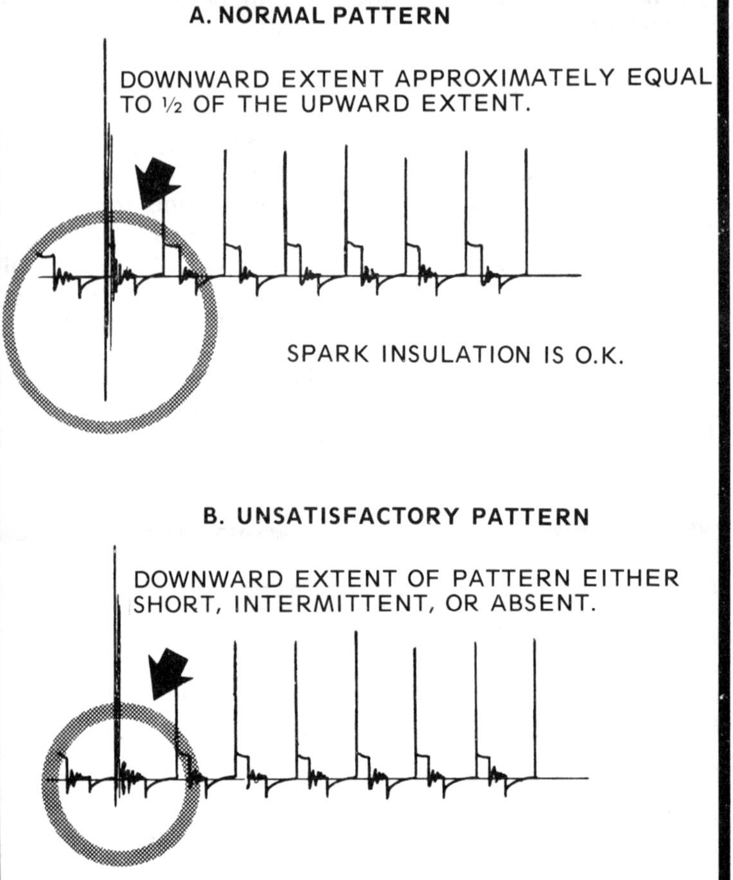

Fig. 8-18. Scope patterns for spark insulation test.

circuited plug should rise to at least 20 KV, Fig. 8-17. If the voltage is under 20 KV check for:

1. Defective coil
2. Low battery voltage
3. Dwell less than specified
4. Excessive resistance in primary circuit due to dirty or faulty battery, cable, and wire connections

Spark Insulation Test

The insulation of all parts of the secondary circuit must be of high quality and in good condition to insure reliability during all operating conditions. Should the insulation of one or more secondary components be defective or have deteriorated to some degree, there may be little or no effect on ignition system operation when the firing voltage requirements of the vehicle are moderate. However, when firing voltage requirements are high, this defective insulation may leak or break down, allowing much of the voltage surge to be lost to ground, resulting in the misfiring of one or more plugs.

The test is performed by using an insulated plier to remove each spark plug wire at the spark plug, in turn, while observing the downward extent of the pattern, see Fig. 8-18. The downward extent should equal at least ½ of the upward extent and should be consistent, not intermittent. If the pattern is unsatisfactory for all cylinders check for:

1. Cracked coil tower
2. Leaky coil wire
3. Cracked rotor
4. Cracked center tower of distributor cap

If one or more cylinders are defective, but not all, check for:

1. Cracked distributor cap towers
2. Leaky spark plug wires

To determine if spark plug wires are at fault, perform the spark insulation test again by removing each wire, in turn, at the distributor cap. If the pattern now shows normal, the spark plug wire is defective.

Spark Voltage Test

Many factors determine spark voltage, such as: spark plug gap, compression pressure, rotor gap, breaks in wiring, and engine speed and load. These factors cause a variation in the spark voltage of an individual plug. However, the spark voltage of all the cylinders should be fairly uniform and within 5 to 15 kilovolts.

The test is performed by comparing the spark voltage of all the cylinders with each other while the engine is running at 1000 rpm, see Fig. 8-19.

If all spark voltages are high, check for:

1. Worn spark plugs
2. Open coil high-tension wire
3. Large rotor gap
4. Defective distributor cap
5. Retarded ignition timing
6. Overall lean mixture

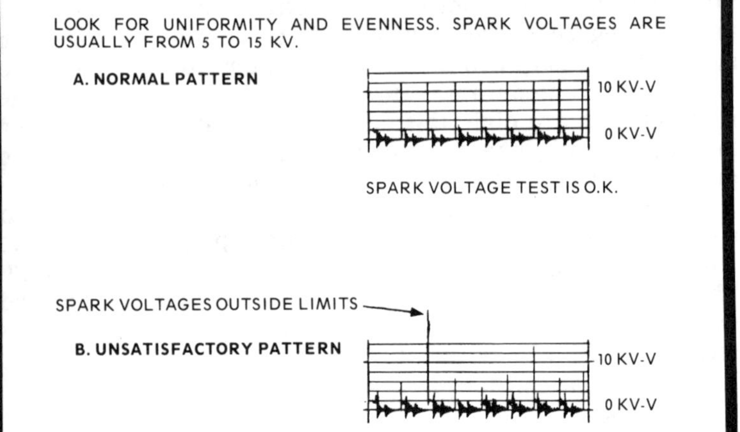

Fig. 8-19. Scope patterns for spark voltage test.

If one or more of the spark voltages are high, check for:

1. Worn spark plug
2. Open spark plug cable
3. Cocked or worn distributor cap
4. Unbalanced fuel mixture (lean)

If all spark voltages are low, check for:

1. Closely gapped spark plugs
2. Advanced ignition timing
3. Overall low compression
4. Overall rich mixture

If one or more of the spark voltages are low, check for:

1. Fouled or closely gapped spark plug
2. Shorted plug cable or cracked insulator
3. Low compression cylinder

Since spark voltage is very important, it is appropriate to explain in detail the reasons for variations in spark voltage. Air (or air-gasoline mixture) is a good electrical insulator and ordinarily offers a very high resistance to the flow of electricity. However, the insulating properties can be destroyed if the air is subjected to a high enough breakdown or spark voltage. When this occurs, the air is ionized, its resistance becomes very low, and a large current flows. Once the air mixture is ionized, very little voltage is required to maintain conduction. The amount of spark voltage required to break down the gap across a spark plug depends on several factors:

Gap Length - The greater the distance separating the electrodes, the greater the spark voltage required to break down the spark plug gap.

Pressure - Greater mixture pressure requires higher spark voltage for breakdown of the spark plug gap.

Temperature - Higher gas or electrode temperatures require less spark voltage.

Composition - Lean mixtures require higher spark voltages than rich mixtures.

Geometry - Sharp edged electrodes require less spark voltage than rounded or worn electrodes.

The scope displays the total voltage drop from the coil tower, through the distributor terminal rotor gap, spark plug leads, and across the spark plug gap to ground. If current is to flow in the secondary circuit, both rotor and plug gaps must conduct at the same time. The total spark voltage required is the sum of the breakdown voltage of each gap and amounts to several thousand volts per gap.

The factors listed above help explain the patterns observed on the scope. For example, a shorted plug requires no breakdown voltage and the scope shows only the spark voltage of the rotor to distributor terminal gap. The spark voltage required for ignition at idle is less than that required at wide open throttle because of cylinder pressure variations. An open high-tension wire means an additional gap which must be broken down before electricity can flow; consequently, several thousand additional volts are required, and a higher signal is seen on the scope screen.

Spark Voltage Under Load

When a load is applied to an engine, the spark voltage rises. This rise will be slight and uniform if the plugs are in good condition and properly gapped. However, if any unusual firing characteristics are observed while the engine is operating under a load, it generally indicates faulty spark plugs. Particularly to be noted are individual firing voltages that are considerably higher or lower than the firing requirements of the other cylinders in the engine.

Opening the throttle rapidly momentarily places a load on the engine. The test is performed by opening the throttle rapidly and observing the scope for any unusual waveforms,

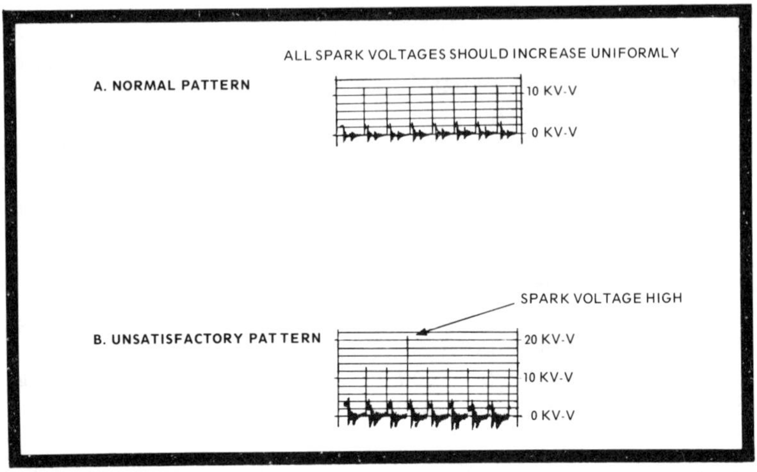

Fig. 8-20. Scope pattern for spark voltage under load test.

Fig. 8-20. It is not necessary to race the engine during this test, since only the short time before the accelerator pump takes effect is of interest.

If all spark voltages are high, check for:

1. Worn spark plugs
2. Open coil high-tension wire
3. Large rotor gap
4. Defective distributor gap
5. Retarded ignition timing
6. Overall lean mixture

If one or more of the spark voltages are high, check for:

1. Worn spark plug
2. Open spark plug cable
3. Cocked or worn distributor cap
4. Unbalanced fuel mixture (lean)

If all spark voltages are low, check for:

1. Closely gapped spark plugs
2. Advanced ignition timing
3. Overall low compression
4. Overall rich mixture

If one or more of the spark voltages are low, check for:

1. Fouled or closely gapped spark plug
2. Shorted plug cable or cracked insulator
3. Low compression cylinder

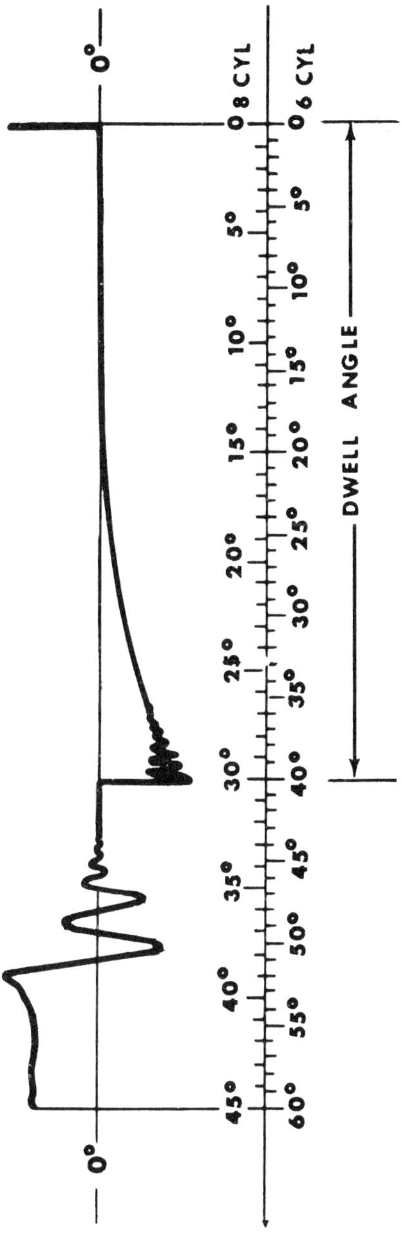

Fig. 8-21. Scope pattern for dwell test.

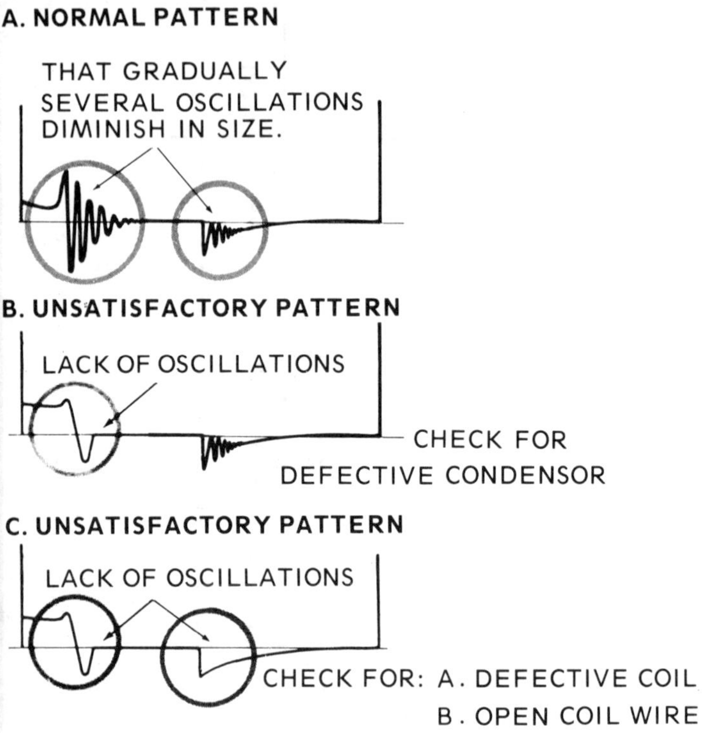

Fig. 8-22. Scope pattern for coil-condenser test.

Dwell Test

While a dwell meter indicates the average dwell for all cylinders, the scope can be used to read the dwell angle for each cylinder. The pattern of one cylinder is expanded until it completely fills the length of the dwell scale, see Fig. 8-21. The correct dwell angle is read directly on the appropriate scale. The dwell angle for any cylinder can be read in a similar manner.

Coil-Condenser Test

A short in the coil or a leaky condenser will cause rapid dissipation of the remaining energy. If not correct the points may be destroyed.

If the coil and condenser are functioning normally, energy remaining in the system after the firing cycle is completed dissipates in diminishing oscillations. The absence of these oscillations is a clear indication of trouble, see Fig. 8-22.

If there is a lack of oscillations in the coil-condenser section, but oscillations exist at the point-close section, check for a defective condenser, see Fig. 8-22, unsatisfactory pattern B.

If, on the other hand, there is a lack of oscillations at both the coil-condenser section and at the point-close signal, check for a defective coil or an open coil wire, see Fig. 8-23, unsatisfactory pattern C.

Point Action Test

Since the points open and close once for each spark plug firing, it is important that they operate properly. The point-action test checks for point arcing, bounce, or poor contact.

The test is performed by expanding the pattern and observing the point-close and point-open signal for each cylinder. The point-close signal should appear as a downward line followed by several oscillations. The point-open signal should appear as an abrupt 90 degree angle, see Fig. 8-23.

If point close signal appears unusual, check for:

1. Point misalignment
2. Point bounce
3. Weak point spring tension

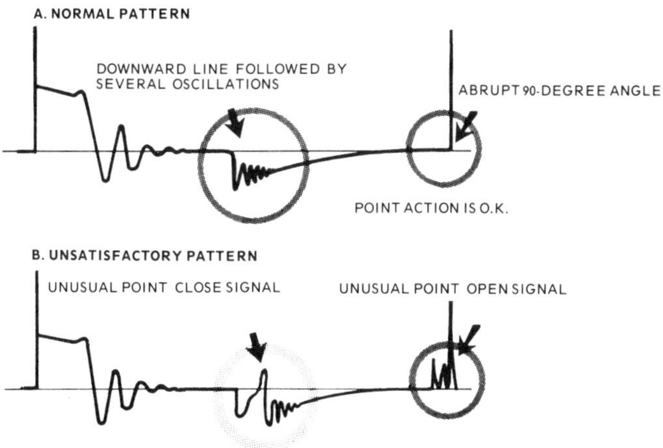

Fig. 8-23. Scope pattern for point action test.

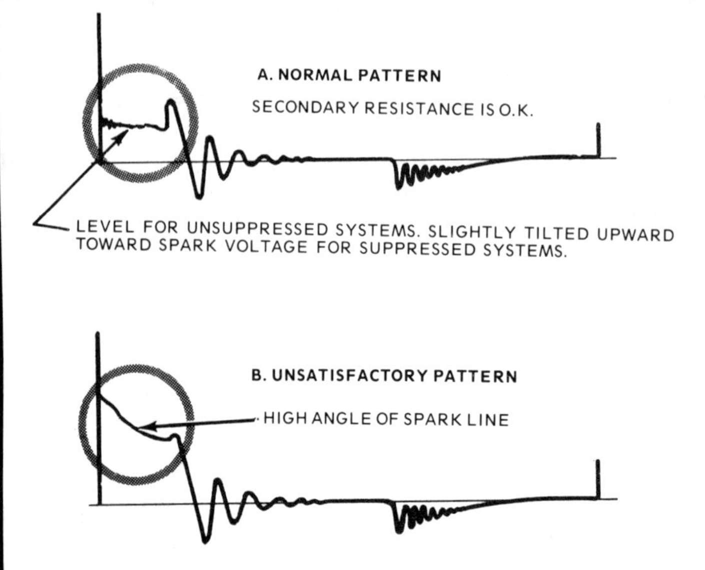

Fig. 8-24. Scope pattern for secondary resistance test.

If point-open signal appears unusual, check for:

1. Burned points
2. Dirty points
3. Faulty condenser

Secondary Resistance Test

Spark line analysis reveals the condition of the secondary circuit conductors. Excessive resistance uses up energy which is needed to maintain good ignition and, consequently, results in poor performance under severe operating conditions.

The test is performed by comparing the spark line of all cylinders for height, length, angle, and oscillations, see Fig. 8-24. The spark lines should be uniform and level for unsuppressed systems and slightly tilted upward toward the spark voltage for suppressed systems.

If the pattern is unsatisfactory and all cylinders are affected, check for:

1. Corroded terminals of coil wire
2. Open coil wire or radio suppressor
3. Burned carbon button of distributor cap

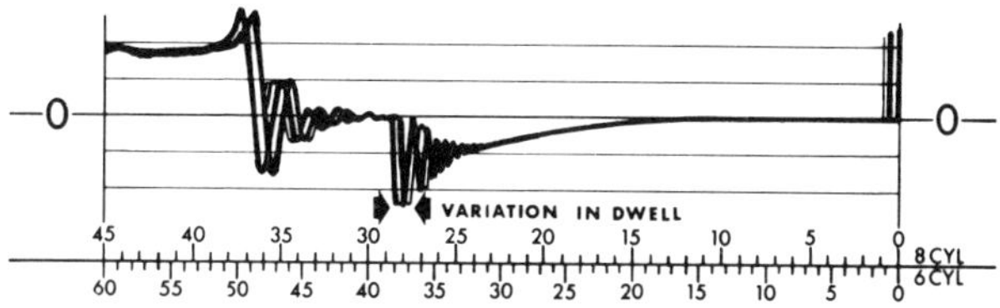

Fig. 8-25. Scope pattern for variation in dwell test.

4. Burned distributor cap segments
5. Burned rotor tip
6. Broken or bent rotor spring

If the pattern is unsatisfactory and one or more, but not all, cylinders are affected, check for:

1. Open spark plug wires
2. Corroded towers of distributor cap
3. Burned distributor cap segments.

Variation-In-Dwell Test

Lowering the superimposed display until the pattern rests on the dwell scale permits any variation in dwell in the distributor cam to be indicated and measured on the dwell scale, see Fig. 8-25. The accuracy of the distributor cam determines the ignition timing relationship of all cylinders. If one or more lobes of the distributor cam become worn, or should the distributor shaft be bent, uneven timing of the cylinders can result.

If variation in dwell is greater than three degrees, for a 6- or 8-cylinder engine or greater than six degrees for a 4-cylinder engine, check for:

Fig. 8-26. Scope pattern of high compression and turbulence.

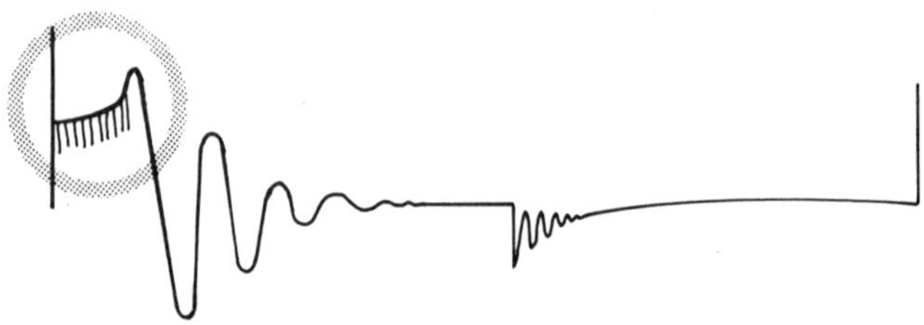

Fig. 8-27. Scope pattern of rich fuel mixture.

1. Worn lobes on the distributor cam
2. Bent distributor shaft
3. Worn distributor bushings

Auxiliary Scope Patterns

The basic scope patterns and tests discussed so far will uncover well over ninety percent of the problems in ignition systems. There are, however, several unusual patterns that one occasionally encounters that must still be discussed. These patterns are the following:

Turbulence - A broken spark line section is caused by high compression and turbulence. It does not necessarily indicate a problem when seen on a high compression engine, but on older cars, see Fig. 8-26, it may indicate excessive carbon deposits or valve malfunction.

Fuel Mixture - This pattern may come and go. It indicates a rich fuel mixture. The carburetor should be checked and adjusted, see Fig. 8-27.

Leaks - If this pattern is seen on a single cylinder, it indicates a manifold leak, see Fig. 8-28. Air is entering the

Fig. 8-28. Scope pattern of air leak.

Fig. 8-29. Scope pattern for unsuppressed system.

combustion chamber and causing an un-mixed fuel mixture condition. When this pattern is seen on all cylinders, a leak at the base of the carburetor is probably the cause.

Unsuppressed Systems - When high tension resistance wire is not used, it is normal for the spark line to contain hash over the entire area, see Fig. 8-29.

Upward Slope of Pattern - An upward slope towards the right can be caused by a wide spark plug gap, high compression, and a lean fuel mixture, see Fig. 8-30.

Open Coil Wire - An open coil winding, open high tension coil cable, or corrosion of the coil tower and cable will present a pattern where all oscillations are broken as shown in Fig. 8-31, or missing entirely.

Faulty Plugs - Shorted and wide spark plug gaps can be detected by close examination of the spark line. A short, high spark line indicates a wide plug gap. A long, low spark line indicates a shorted plug, see Fig. 8-32.

Auxiliary Gap Spark Plugs - Auxiliary gap spark plugs cause the tail of the first oscillation to alternately vary in length, see Fig. 8-33.

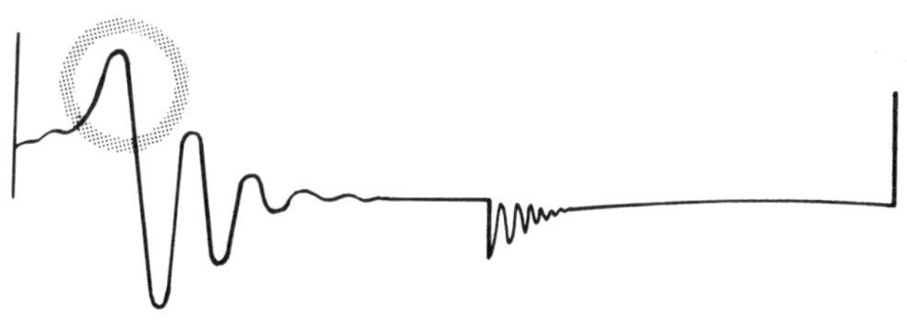

Fig. 8-30. Scope pattern caused by wide plug gap, high compression, or lean fuel mixture.

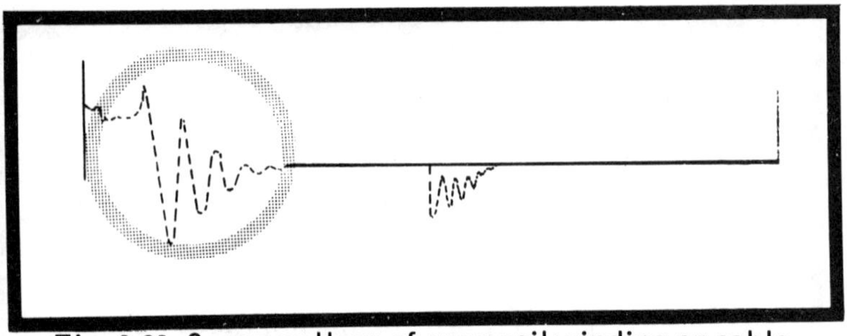

Fig. 8-31. Scope pattern of open coil winding or cable.

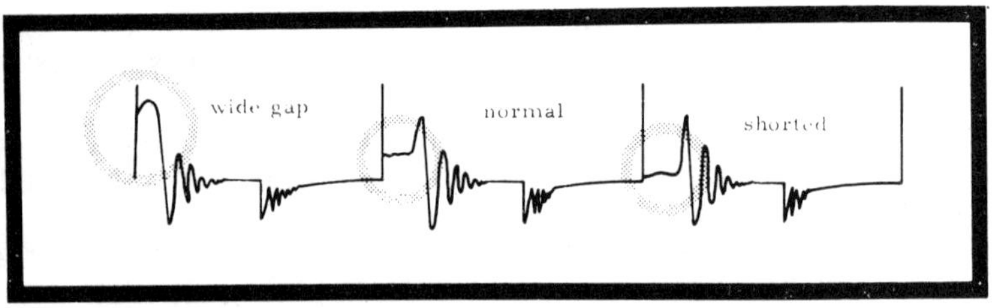

Fig. 8-32. Scope pattern of faulty plugs.

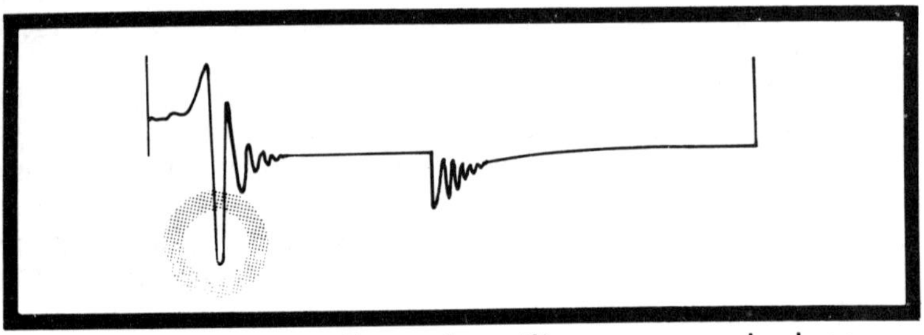

Fig. 8-33. Scope pattern of auxiliary gap spark plugs.

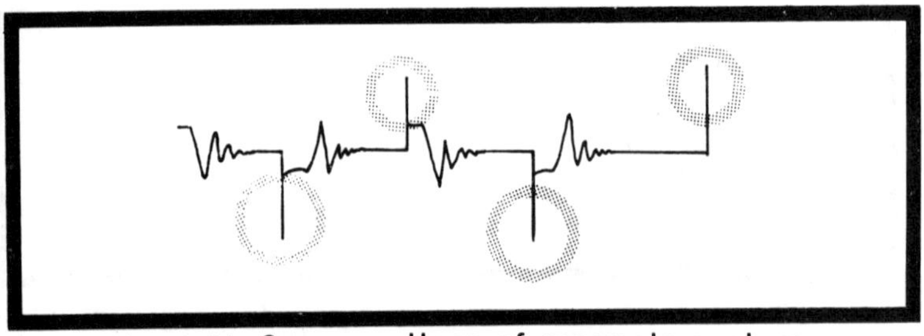

Fig. 8-34. Scope pattern of magneto system.

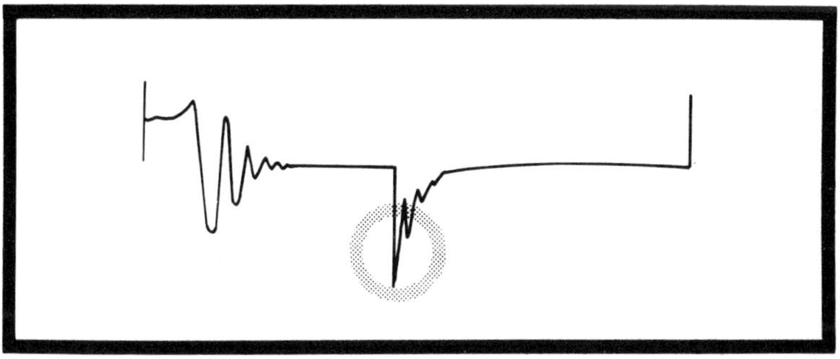

Fig. 8-35. Scope pattern of transistorized ignition system.

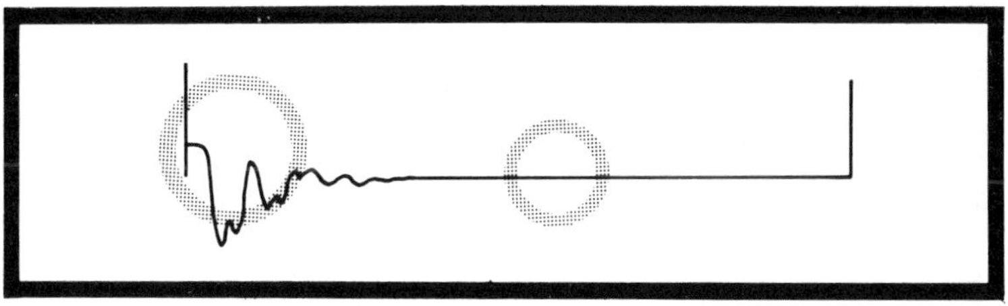

Fig. 8-36. Scope pattern of capacitive discharge system.

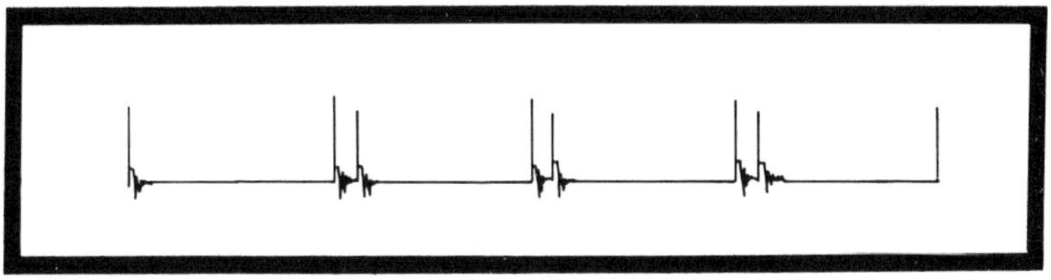

Fig. 8-37. Scope pattern of dual-spark capacitive discharge system.

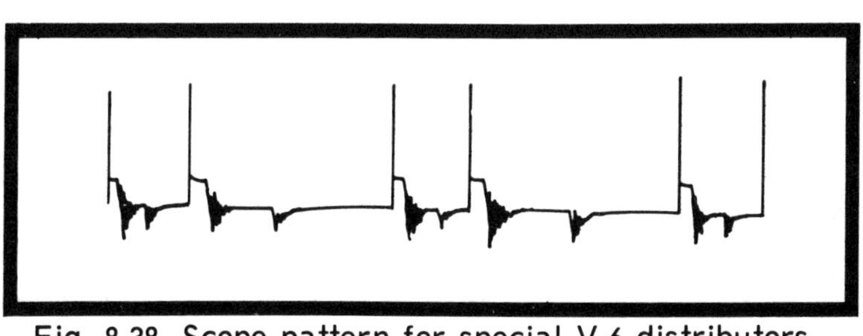

Fig. 8-38. Scope pattern for special V-6 distributors.

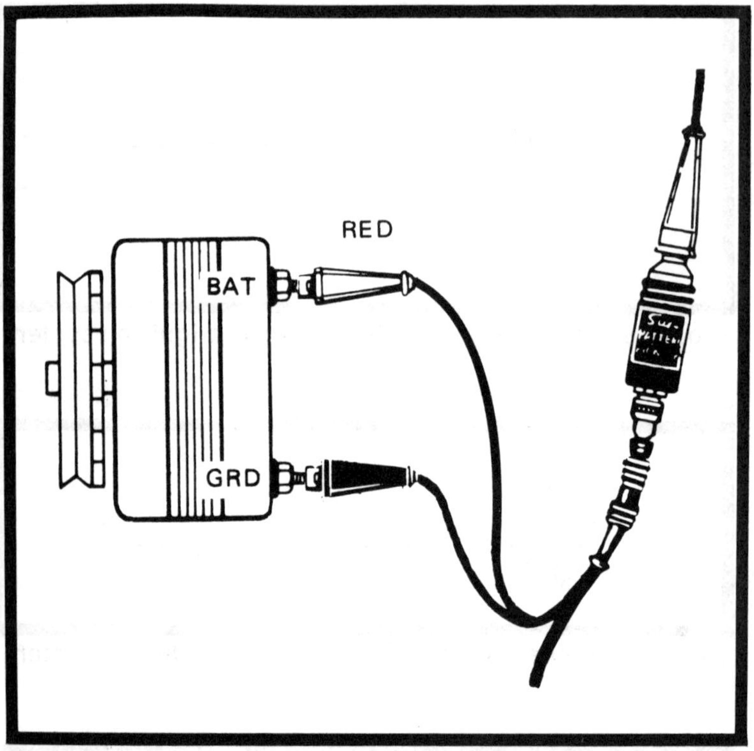

Fig. 8-39. Scope leads connected for alternator diode test.

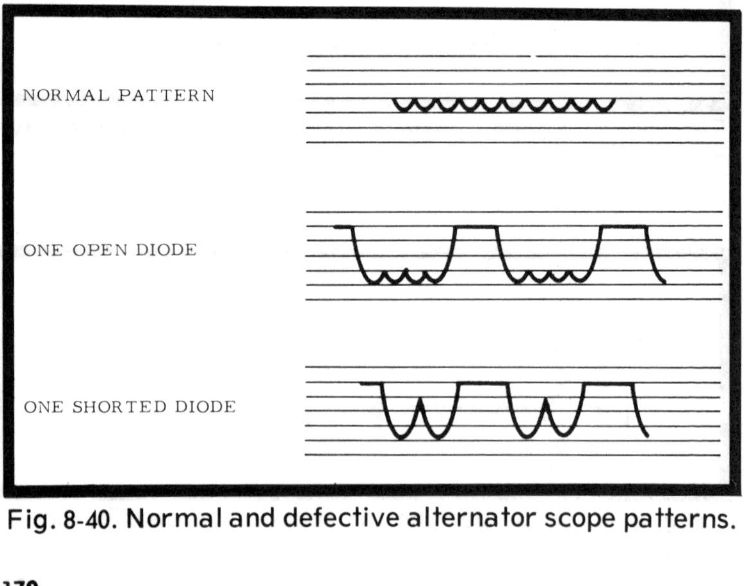

Fig. 8-40. Normal and defective alternator scope patterns.

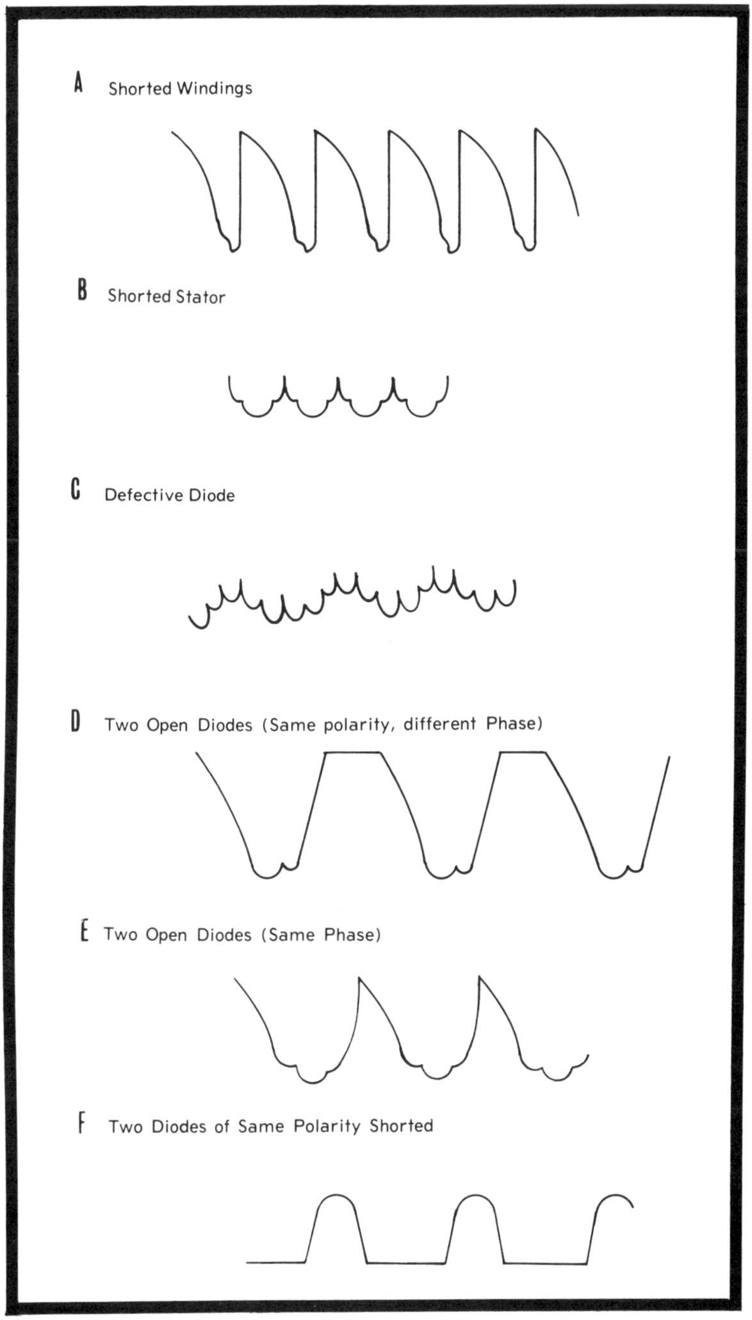

Fig. 8-41. Additional defective alternator scope patterns.

Magneto System - A magneto system pattern alternates polarity with each firing of a spark plug, see Fig. 8-34.

Transistorized Ignition Systems - The pattern for a transistorized ignition system is similar to the conventional pattern except that the wave form deflects farther below the zero line at the point-close signal and the oscillations diminish more rapidly, see Fig. 8-35.

Capacitive Discharge Systems - A capacitive discharge system is characterized by the lack of a point-close signal and a very short spark line, see Fig. 8-36. The dual-spark Delco Remy capacitive discharge system is shown in Fig. 8-37. This system fires each spark plug twice.

V-6 Engines - Special V-6 distributors used on Buick V-6 engines have a cam lobe spacing of 75 and 45 degrees. This produces alternately narrow and wide-spaced patterns in the normal pattern, see Fig. 8-38.

Alternator Diode-Stator Test

The scope can also be used for checking alternator diodes and stator. The alternator need not be removed from the engine for this test, since the test is performed with the alternator operating under normal conditions.

On some scopes, the primary leads are connected as shown in Fig. 8-39. One lead is connected to ground and the other is connected to the alternator output (BAT) terminal. Other scopes come supplied with a special alternator lead which is connected to the battery.

The pattern for a normal system is shown in Fig. 8-40. Note that the signal is small and uniform. A defect in the system causes a larger pattern with unusual lumps, (compare patterns in Fig. 8-40). There is generally little difficulty in detecting a fault in the system with the scope, so that it is very useful as a quick test on the alternator. Additional defective alternator patterns are illustrated in Fig. 8-41.

Chapter 9

Electronic Compression Testers

The electronic compression tester enables the mechanic to make fast comparison compression tests without removing spark plugs. An electronic circuit momentarily shorts any selected cylinder so that it will not fire. When the cylinder is shorted, the engine rpm drops, and the drop in rpm is registered on a meter. If each cylinder's compression and firing condition is normal, the meter indications for all cylinders will be the same. A cylinder which does not contribute to the total load will not affect the rpm when shorted. and, therefore, little or no change in reading will be observed.

Obviously, if an electronic compression test were made without first analyzing the electrical components of the ignition system, the results would be inconclusive. All the electrical components of the ignition system should be known to be in good working order so that the only condition which will determine power loss will be the mechanical state of the engine.

Various names have been given to the electronic compression test by the manufacturers of test equipment. Some of these names are as follows:

Power Balance Test
Dynamic Compression
Power-Check Compression Test
Electro-Power Balance Test

The basic concept behind the test is the same, regardless of what the test is called.

The test can be performed manually by simply shorting out one plug at a time with a jumper wire and measuring the drop in rpm with a tachometer. The manual procedure, however, is difficult in application and very time consuming. Because of this, the electronic devices have achieved popularity.

In addition to the electronic compression test, a carburetion balance test can be performed by many of these testers.

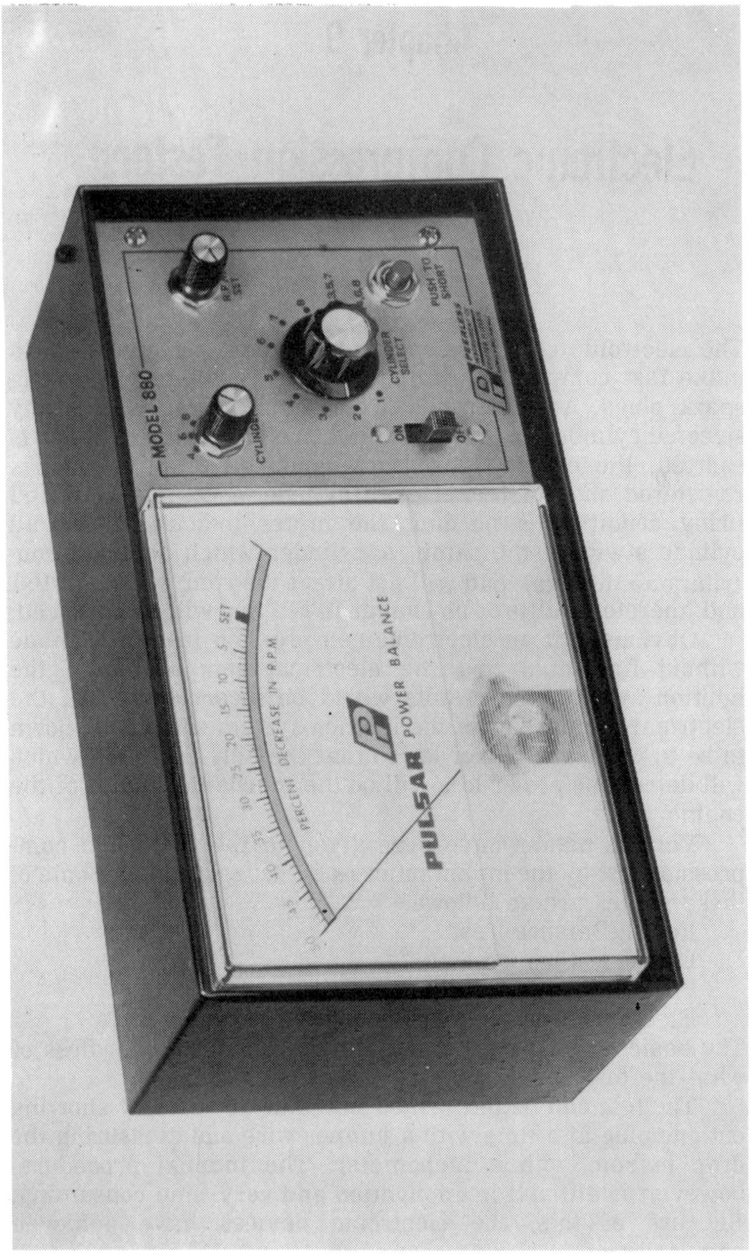

Fig. 9-1. Single unit electronic compression tester.

TYPES OF TESTERS

There are two types of electronic compression testers: one type is a unit complete in itself; the other is a unit used in conjunction with an oscilloscope.

Fig. 9-1 is an example of a complete unit. A cylinder switch is rotated to either 4, 6, or 8, depending on the number of cylinders in the engine under test. The rpm set control calibrates the instrument to the set bar on the meter for any test speed. A cylinder select switch selects the cylinder that is to be shorted, and the push-to-short button allows shorting of the selected cylinder only when the button is depressed.

Fig. 9-2 is an example of an instrument that is used in conjunction with an oscilloscope. In place of the cylinder select switch, there is a cylinder selector control which is slowly rotated while observing the oscilloscope. The oscilloscope pattern shows which cylinder is shorted. In Fig. 9-3 the cylinder selector control has been set so that the fifth cylinder in the firing order is shorted.

The pattern indication which shows the shorted cylinder is the notch or down-step shown in Fig. 9-3. As the cylinder selector control is turned the pattern indicator moves across the screen. As the pattern indicator passes through each cylinder's pattern, the plug voltage spike disappears and that cylinder misses or is shorted. A disadvantage of this type of

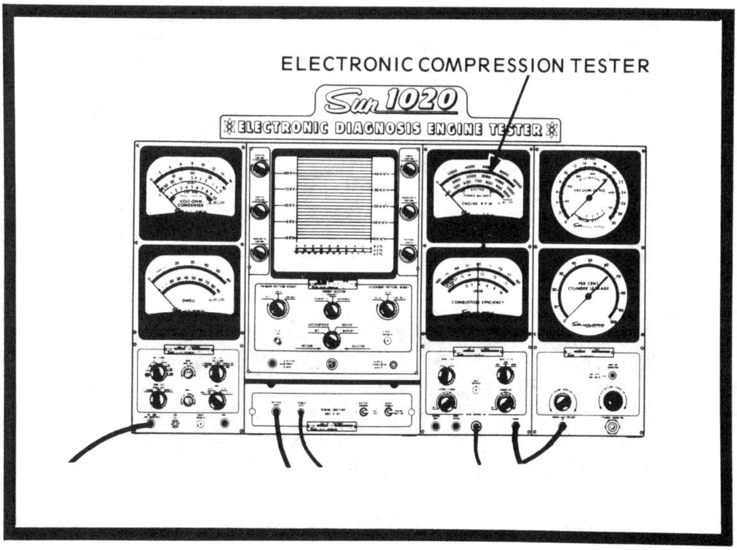

Fig. 9-2. Electronic compression tester for use with scope.

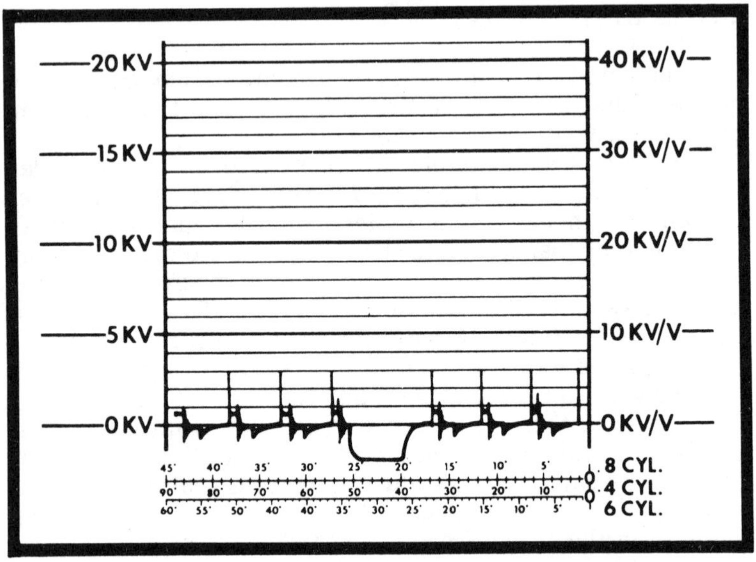

Fig. 9-3. Scope pattern with fifth cylinder shorted.

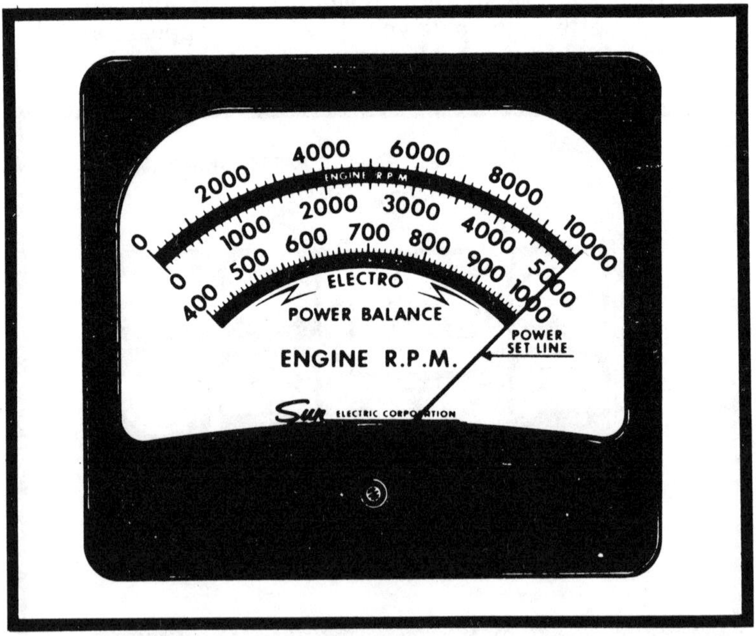

Fig. 9-4. Tachometer used for electronic compression test.

tester is that when number one cylinder is shorted, the scope loses its trigger signal and the pattern becomes free running.

Scales

When a spark plug is electronically shorted, the engine rpm drops. The most straight-forward way in making a measurement of this drop is to read it in rpm from a given set rpm on a tachometer. Fig. 9-4 shows an expanded-scale tachometer on which the set speed is 1000 rpm. An expanded scale is used so that the drop in rpm will result in a greater meter swing giving greater accuracy. An ordinary tachometer can be used, but the total meter swing may only be about 1/8 inch.

Frequently, the scales will be marked in percentage. Fig. 9-5 shows a scale on which readings are made as a percentage decrease in rpm from the set rpm. Other scales, such as that shown in Fig. 9-6 will show the decrease in rpm as an upward reading either in percentage or in rpm. Some instruments using a percentage scale will have a meter set control on the

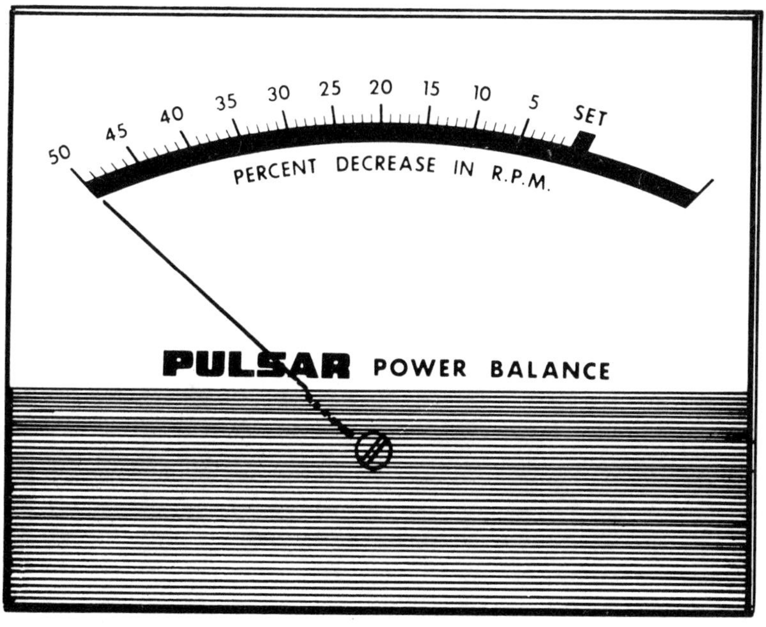

Fig. 9-5. Downward reading percent scale.

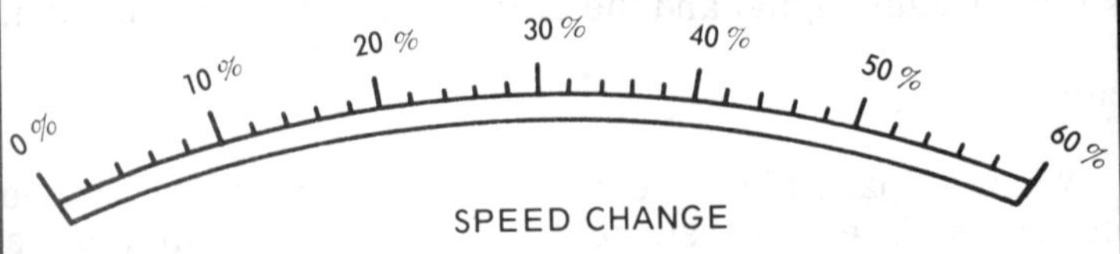

Fig. 9-6. Upward reading percent scale.

front panel which will allow the compression test to be made at any rpm from idle speed to beyond 2000 rpm.

All of these scales do basically the same thing. That is, they all indicate the change in rpm from the set rpm whenever a plug is electrically shorted.

HOOK-UP

Three lead connections, as shown in Fig. 9-7, are needed to hook-up the electronic compression tester. One lead is con-

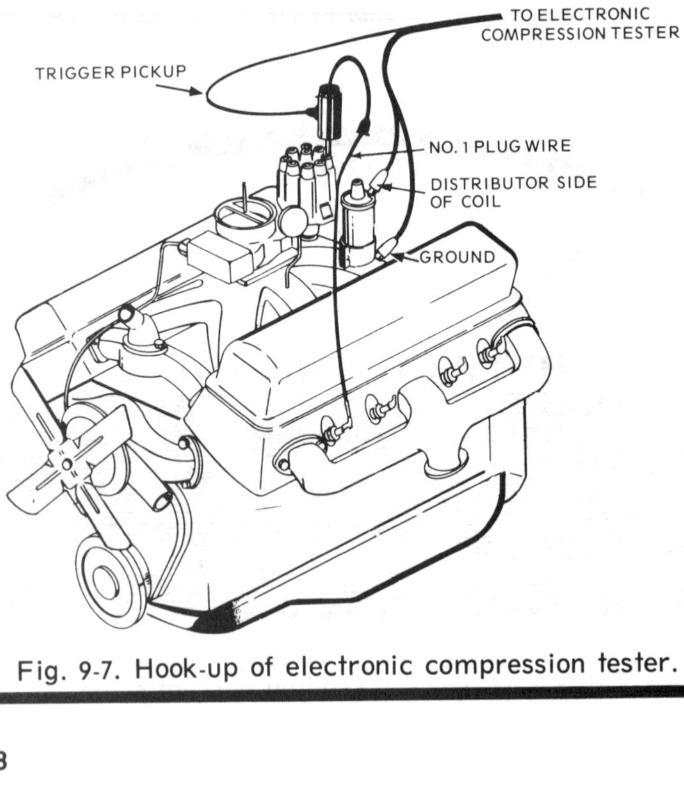

Fig. 9-7. Hook-up of electronic compression tester.

nected to the distributor side of the coil. It is through this lead that the ignition system on the car is shorted by the tester, thereby preventing the coil from delivering a voltage pulse which would ordinarily fire the spark plug.

The second lead is the trigger pickup, which is connected to number one spark plug. This lead introduces a voltage pulse into the electronic compression tester so that the circuit within the tester may be synchronized with engine speed.

The third lead is connected to a good ground and is used to complete the electrical circuit from the engine to the tester.

Electronic Compression Balance Test

The purpose of the electronic compression balance test is to determine if all cylinders are contributing their share to the overall operating efficiency of the engine. Any condition which will cause one cylinder to become ineffective, even though it continues to fire, will decrease the overall efficiency of the system. After the tester has been connected to the engine, the cylinder switch must be set to the 4, 6, or 8 position corresponding to the number of cylinders in the vehicle under test. The engine is started and adjusted to the test speed, which is ordinarily approximately 1000 rpm.

The cylinder selector switch is set to the first position. This switch is so designed that it will short the cylinders of an engine in a numerical sequence corresponding to the particular firing order of the engine under test. For example, on an engine having a firing order of 18436572, the first position of the switch selects the No. 1 cylinder, the second position of the switch selects the No. 8 cylinder, the third position of the switch selects the No. 4 cylinder, etc. All of the cylinders are tested in order and the meter indications are recorded for each cylinder.

Good cylinders developing an equal amount of power will have approximately the following percent decrease in rpm:

8 cylinders - 5 percent to 8 percent
6 cylinders - 7 percent to 10 percent
4 cylinders - 10 percent to 13 percent

The actual readings may vary with different engines. The readings, however, for each engine should be uniform.

If the meter pointer remains at the set point when a cylinder is shorted, that particular cylinder is dead. It produces no power, even when the plug fires.

A reading showing less than normal decrease indicates low compression due to worn rings, burned or sticky valves,

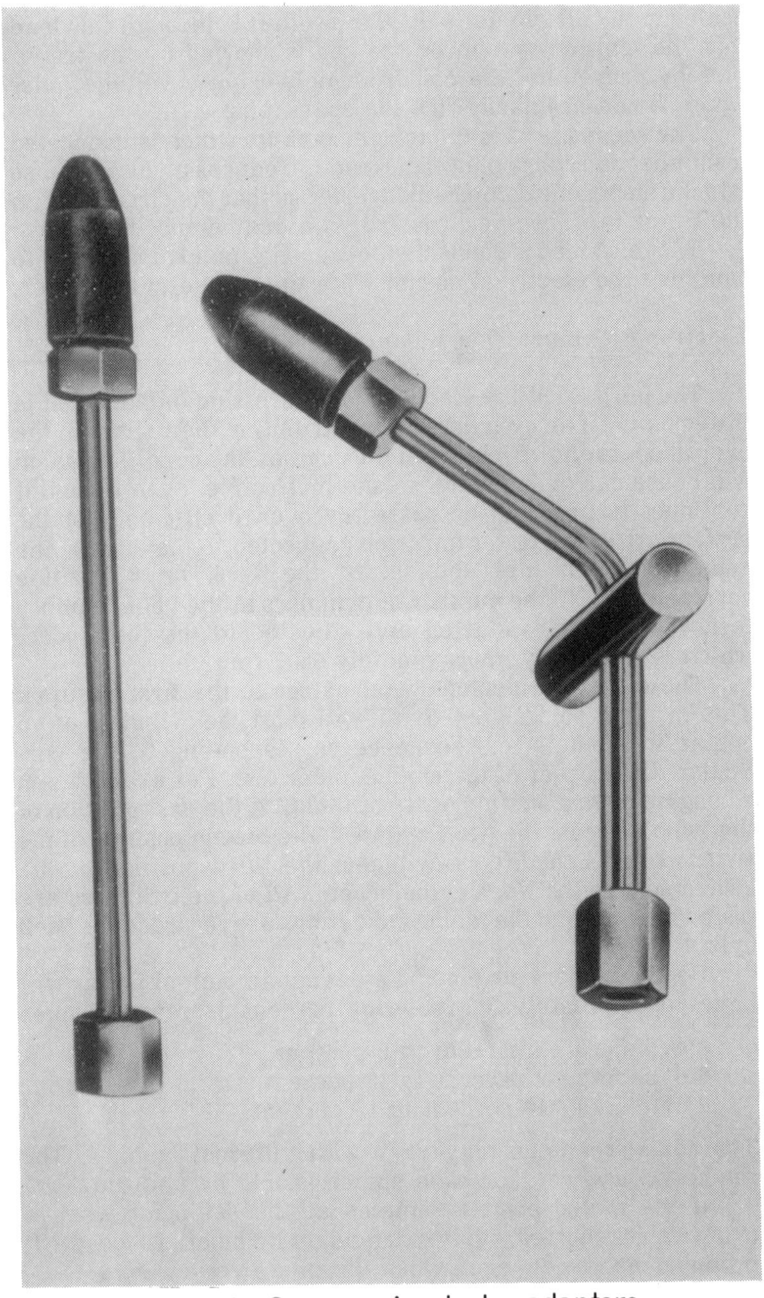

Fig. 9-8A. Compression tester adapters.

Fig. 9-8B. Compression tester.

defective valve lifters, leaky manifold, or any other factor contributing to power loss in the cylinder. A faulty ignition system can also cause a reading less than normal, so that it is important to know the condition of the ignition system before this test is made.

If all readings follow a pattern of high, low, high, low, etc., the carburetion balance test should be performed before condemning any cylinders.

Carburetion Balance Test

The carburetion balance test provides a quick means of testing the idle mixture screw adjustments of dual or four-throat carburetors. The test is made by comparing the results of shorting the cylinders fed by one throat of the carburetor with the results of shorting the cylinders fed by the other throat.

The test is ordinarily performed only on V-block engines. Referring to the firing order of 18436573, when the cylinder select switch is set in one position, the first, fourth, sixth, and seventh cylinders of the engine are shorted out at one time. The other position of the cylinder selector switch shorts out the eighth, third, fifth, and second cylinders of the engine.

The test is performed at approximately 1000 rpm. If there is more than a 3 percent difference between the two readings, balance the idle mixture according to manufacturer's procedure, and retest. If there is still more than a 3 percent difference, check for vacuum leaks, defective vacuum operated accessories, dirty carburetor, leaking or cracked intake manifold gaskets, and intake manifold restrictions.

COMPRESSION TESTER

When a problem in an engine is uncovered by using the electronic compression tester, it is generally a good practice to use an ordinary compression tester to verify the results and to pin-point the problem more accurately. A compression tester along with several long-reach adapters is shown in Fig. 9-8.

After having run the engine long enough to bring it up to operating temperature, stop the engine and remove all spark plugs and gaskets. Then set the carburetor throttle valve wide open and ground the high-tension coil wire.

Hold the compression tester firmly in a spark plug hole with enough pressure for the rubber cone on the adapter to seal the hole, see Fig. 9-9. Crank the engine until no increase in pressure is noted on the gauge; normally 3 to 4 revolutions is enough. Record this reading and repeat the test for all the cylinders in the engine.

The actual specification in pounds is not of utmost importance. It is more important that all cylinders have approximately the same compression pressures. On high-compression engines (150 lbs and higher) the pressure should not vary more than 15 lbs from the highest to the lowest cylinder. On low-compression engines (under 150 lbs) the pressure should not vary more than 10 lbs from the highest to the lowest cylinder. If a greater than permissible variation of pressure exists between cylinders, a degree of engine roughness will exist that a tune-up cannot correct.

A high-compression reading is caused by an accumulation of carbon in that cylinder. High-compression readings in all cylinders is caused by an even accumulation of carbon in all cylinders. This may cause pinging that is difficult to eliminate.

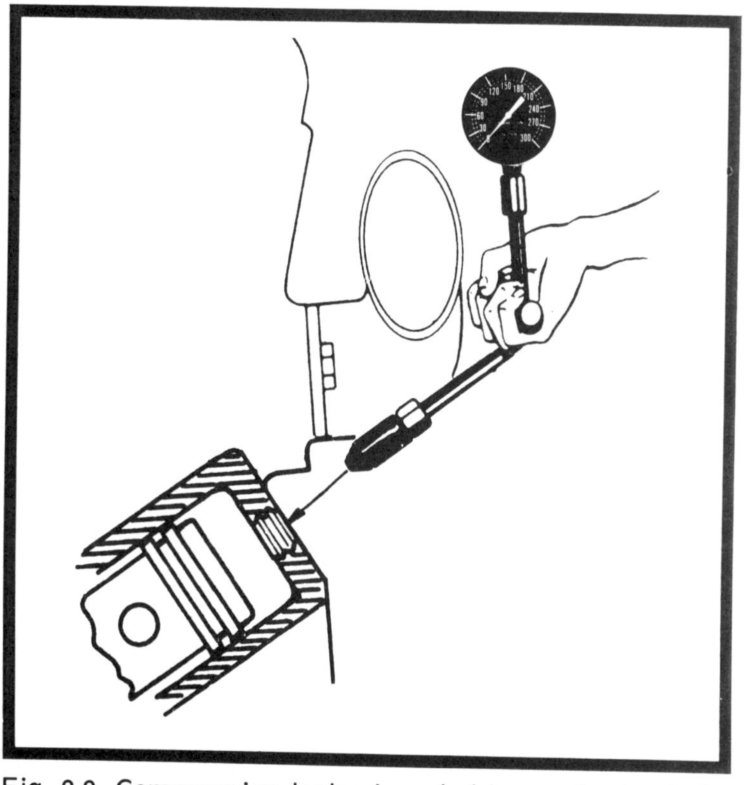

Fig. 9-9. Compression tester inserted in spark plug hole.

A low-compression reading is caused by leakage through:

1. Piston rings that are worn. If this is suspected, pour a teaspoon of oil into the spark plug hole to seal the rings, and recheck compression. If compression reading increases, the rings are bad.

2. Valves that are not seating properly. The addition of oil will not materially affect the readings obtained.

3. Head gasket. An additional indication of this particular trouble is the appearance of water in the crankcase.

Chapter 10

Vacuum Gauges

Vacuum results from a difference in pressures, when one of the pressures is below atmospheric pressure. The normal or standard atmospheric pressure at sea level is 14.7 pounds per square inch (psi) or 29.92 inches of mercury. The chemical symbol for mercury is Hg, and this is commonly used in stating the amount or degree of vacuum. For example, 19 in. Hg means nineteen inches of mercury.

The usual vacuum gauge used in automotive tune-up work is made so that, in addition to giving vacuum readings in inches of Hg, it will also indicate pressures up to 10 psi above atmospheric pressure. This is so that fuel pump pressures can be measured.

A simple and convenient vacuum gauge is shown in Fig. 10-1. Vacuum readings from 0 to 30 inches of Hg are read from left to right on the outer scale, and pressure readings from 0 to 10 psi are read from right to left on the inner scale.

HOW THE VACUUM GAUGE WORKS

The operation of the vacuum gauge is based on a Bourdon tube. This is a tube which is flattened so its cross section is an oval, and which is bent into a circular form with the flat side out, see Fig. 10-2. The tube looks like about two thirds of the circumference of a circle. One end is anchored in the instrument and the other is free to move. The free end is connected to linkage and gears to make the indicating hand move. When pressure is applied to the inside of the Bourdon tube, it tries to straighten out in proportion to the pressure applied, so that the outer end of the tube moves outward. This movement in turn moves the indicating pointer so that it points to the figure representing the amount of applied pressure. When vacuum is applied, the pressure inside the tube is lowered, and atmospheric pressure tries to make the tube bend into a smaller circle. The end of the tube moves inward in proportion to the amount of vacuum and indicates accordingly.

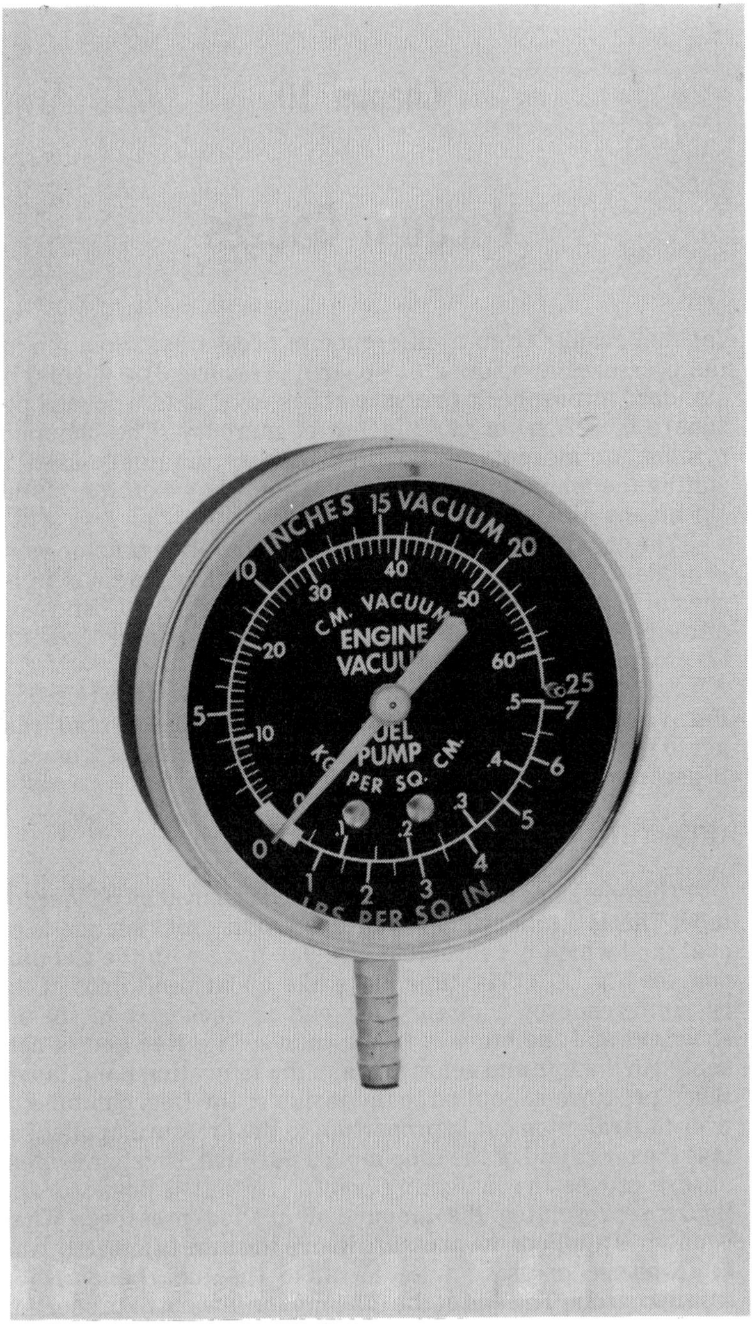

Fig. 10-1. Typical vacuum gauge.

To interpret vacuum gauge indications, it will help to know the causes of vacuums in an engine that are checked with the vacuum gauge. On each intake stroke, the piston moves downwards (towards the crankshaft), decreasing the pressure in its cylinder and in the intake manifold. In effect, it is pulling the air past the throttle valve, through the intake manifold, and past the intake valve. At idle speed with the throttle slightly open, only a small amount of air can get by. Yet the piston moves down its full stroke, so a volume much greater than the entering volume of air at atmospheric pressure is displaced by the piston. This causes the air to be "stretched," so to speak, and its pressure is lowered: in other words, vacuum is created. The amount of vacuum will be in proportion to throttle opening and engine speed, the tightness of the induction system, condition of the rings, pistions, and valves.

The amount of vacuum will change with every change in engine speed and throttle opening, and also with changes in the engine's mechanical condition. Therefore, by observing the instrument readings under different operating conditions, a good idea of the condition of the engine can be obtained. Also, the readings will give helpful information for making certain adjustments.

Cranking Vacuum Test

A high steady vacuum during cranking indicates a mechanically sound engine, properly sealed intake manifold

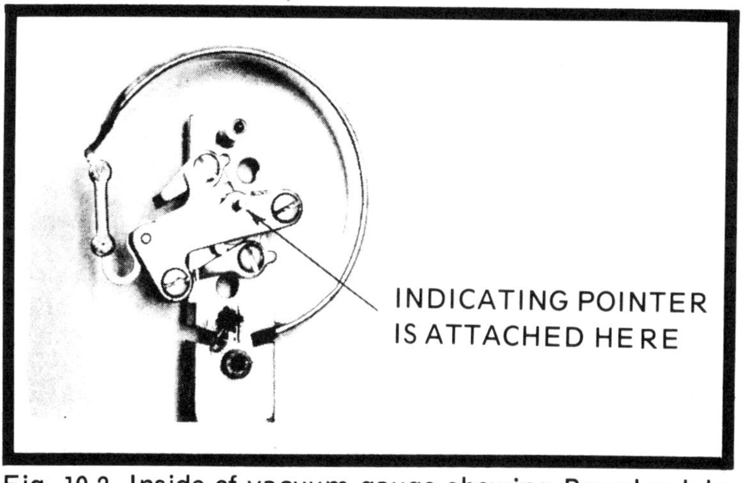

Fig. 10-2. Inside of vacuum gauge showing Bourdon tube and linkage.

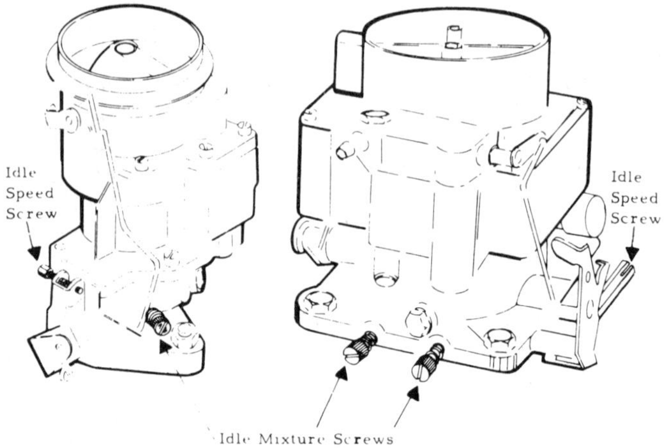

Fig. 10-3. Location of idle mixture screws.

and intake valve guides. Even though the individual design characteristics of different engines cause the normal cranking vacuum to vary considerably from one type to another, the results of this test have been found to be extremely valuable as a guide in determining compression factor conditions and intake system leakage.

The cranking vacuum test is performed with the engine at normal operating temperature, but not running. Remove the high-tension coil lead from the distributor cap, and ground it. Then remove the air filter and back out the idle speed screw, Fig. 10-3, until the throttle valve is tightly closed. If the carburetor is equipped with an idle speed air screw, turn the screw in until it bottoms lightly. In both cases count the number of turns so that the screws can be returned to their original positions after the tests.

If the engine is equipped with a PCV (Positive Crankcase Ventilation) system, remove the PCV valve at the engine rocker-arm cover or push-rod cover, see Fig. 10-4, and plug the valve with a piece of tape. Using the hose supplied, connect the vacuum guage to a source of manifold vacuum. This may be a fitting on the carburetor below the throttle plate, or a provision may be made to install a fitting into the intake manifold. Crank the engine and note the vacuum gauge reading.

General engine condition is indicated from one of three possible gauge readings:

1. A steady vacuum reading of 6 inches or more indicates a mechanically sound engine, see Fig. 10-5. The actual reading varies from one engine to another.

2. An excessively low steady vacuum reading is caused by a condition which affects all cylinders equally, see Fig. 10-6. This can be caused by the following:

(a) Leaking carburetor flange gasket
(b) Worn carburetor throttle shaft
(c) Leaking vacuum lines
(d) Improper valve timing
(e) Slow engine cranking due to a run down battery, defective starter motor, or excessive mechanical drag in engine.

3. A pointer which pulses regularly indicates a leaky condition which affects one particular cylinder consistently, see Fig. 10-7. A larger dip indicates a larger leak, check for the following:

(a) Burned or stuck valve
(b) Intake manifold leak at one cylinder

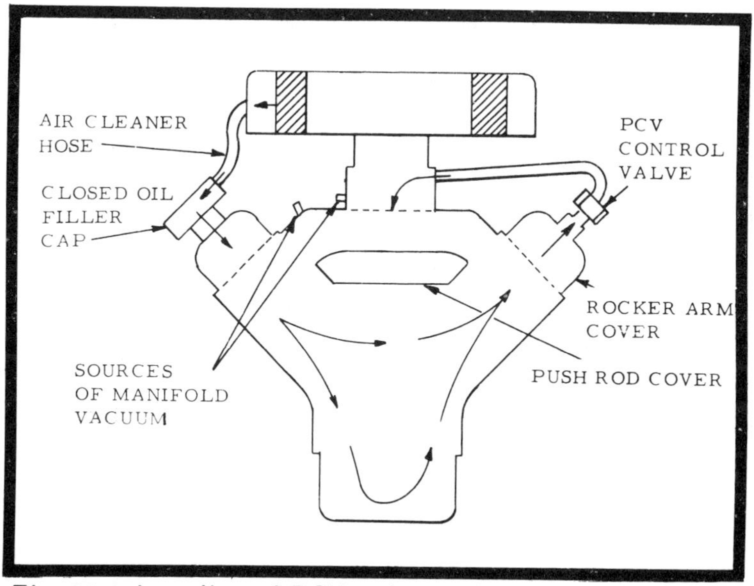

Fig. 10-4. Location of PCV valve and sources of manifold vacuum.

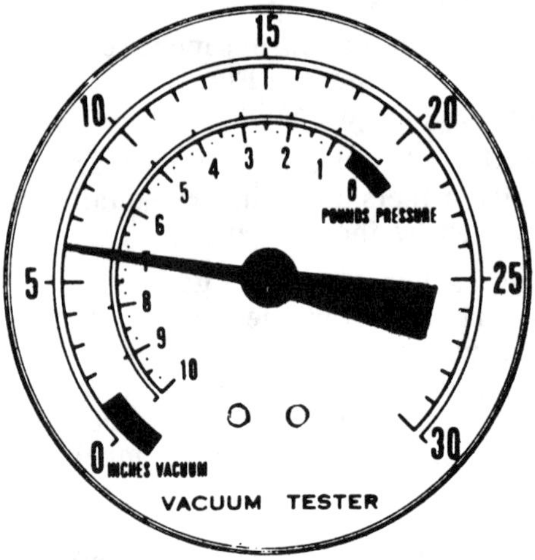

Fig. 10-5. Steady vacuum gauge reading for cranking vacuum test.

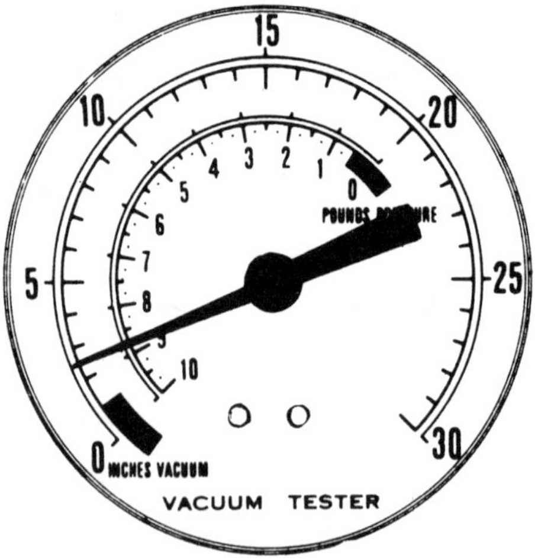

Fig. 10-6. Low vacuum gauge reading.

(c) Worn intake valve guide
(d) Broken rings or piston

Unplug the PCV valve (previously plugged with a piece of tape), and crank the engine again. If the PCV valve is operating properly, the vacuum will drop to about ½ the value noted earlier. A reading much lower than ½ indicates excessive flow which could upset the proper carburetor air-fuel ratio and cause rough idling and burned valves. No change in the vacuum indicates a clogged PCV valve.

Running Vacuum Test

The running vacuum test is performed by connecting the vacuum gauge to a source of manifold vacuum, and operating the engine at normal temperature and idle speed. Engine condition is indicated by the following vacuum guage readings:

1. A steady reading of between 15 and 22 indicates a mechanically sound engine, see Fig. 10-8.
2. A pointer which sweeps or wanders erratically through several inches indicates a malfunction affecting all cylinders

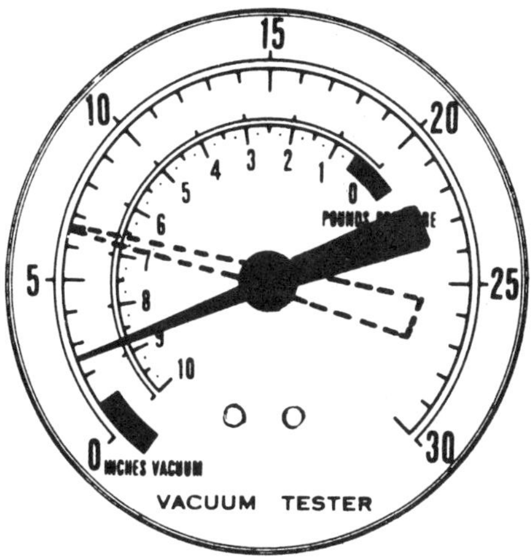

Fig. 10-7. Pulsating vacuum gauge reading.

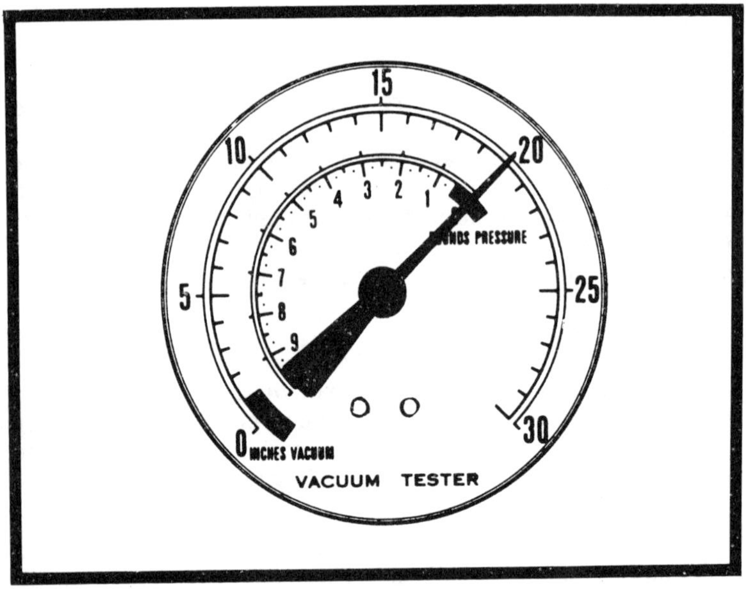

Fig. 10-8. Steady vacuum gauge reading for running vacuum test.

unequally and not consistantly, see Fig. 10-9. To help isolate the trouble area, run the engine at about 2000 rpm. If the pointer steadies, check the following:

(a) Ignition and timing
(b) Centrifugal advance, (loose springs at idle)
(c) Carburetor mixture adjustment at idle

If the sweeps get larger, check for weak or broken valve springs. If the sweeps become shorter and more rapid, check for intake system leaks or sticky valves.

PCV System Test

On engines equipped with a PCV system, remove the air cleaner hose from the rocker-arm cover or closed oil filler, see Fig. 10-4. Hold the vacuum gauge rubber hose fitting firmly over the engine opening.

A properly working PCV system will draw a vacuum of about 3 to 5 inches within 10 seconds with the engine operating at idle speed. If there is very little change in gauge reading, the PCV valve is clogged or there are excessive air leaks into the crankcase.

Exhaust Restriction Test

This test is used to determine whether the exhaust system is restricted, or offers excessive back pressure which could result in reduced engine efficiency.

With the vacuum gauge connected to a source of manifold vacuum, increase engine speed to 2000 rpm, maintain this speed for approximately 10 seconds, and note the vacuum gauge reading. If the exhaust system is restricted, the vacuum reading will gradually decrease.

Fuel Pump Pressure and Volume Test

In order to assure an adequate quantity of fuel and constant fuel level for all driving conditions, it is necessary that both fuel pump pressure and volume be within specifications. To perform the pressure test, disconnect the fuel line at the carburetor and attach a vacuum gauge with a pounds pressure scale to the fuel line, see Fig. 10-10. Operate the engine at idle speed, and, with the gauge held at carburetor height, note the reading.

Compare the reading with manufacturer's specifications. If specifications are not available, fuel pump pressure can be

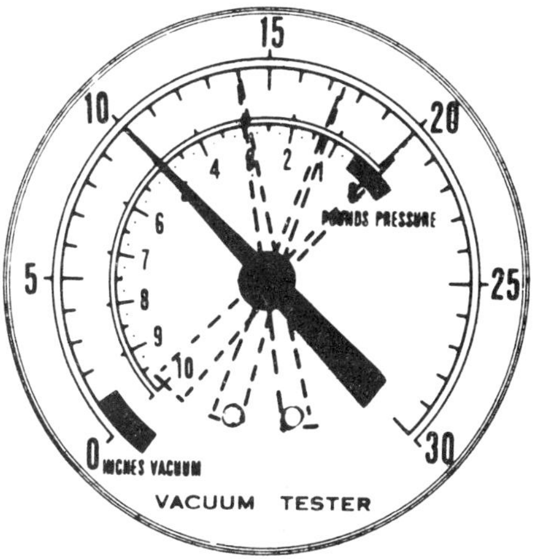

Fig. 10-9. Wandering vacuum gauge reading.

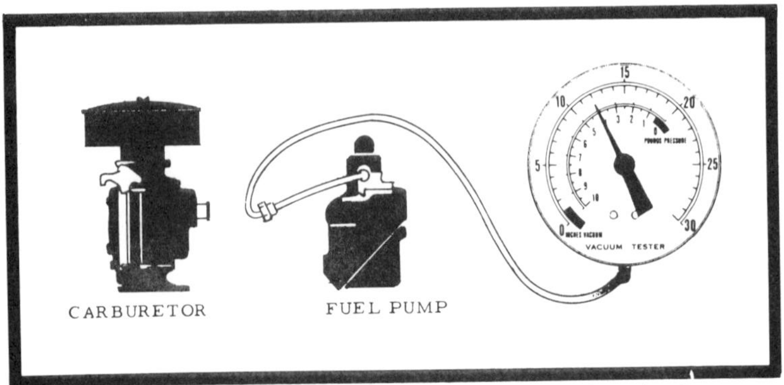

Fig. 10-10. Connection of vacuum gauge for fuel pump pressure test.

considered satisfactory if it is between 4 and 6 psi, with lower readings for smaller displacement engines. If the pressure reading falls outside this range, correct specifications should be obtained before the fuel pump is condemned.

To perform the volume test, temporarily reconnect the fuel line and run the engine to fill the carburetor fuel bowl, then stop the engine. Disconnect the fuel line at the carburetor again, and connect a flexible hose to the line. Lead the hose to a container marked in fluid ounces, see Fig. 10-11.

Operate the engine and collect the discharged gasoline from the fuel pump for 30 seconds. Consult manufacturer's specifications for the exact delivery rate, or be guided by the table shown in Fig. 10-12. Insufficient volume with normal pressure indicates a restricted or leaky fuel line or a defective fuel pump.

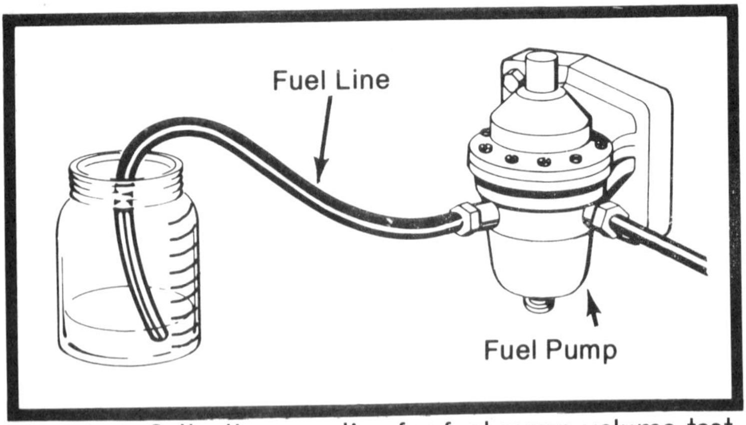

Fig. 10-11. Collecting gasoline for fuel pump volume test.

ENGINE DISPLACEMENT IN CUBIC INCHES	OUNCES COLLECTED IN 30 SECONDS
Up to 225	8
225 to 350	11
Over 350	16

Fig. 10-12. Amount of gasoline discharge for fuel pump volume test.

Fuel Pump Vacuum Test

The fuel pump vacuum test is an auxiliary test and is necessary only when the pump output does not meet the pressure or volume specifications. The fuel pump vacuum test is made to determine whether the defect is in the pump or in the fuel line.

To perform the test, disconnect the fuel pump flexible line from the fuel tank delivery line. With the proper adapter, attach the vacuum gauge, see Fig. 10-13. Start the engine and operate it at idle speed for a period of time sufficient to permit the vacuum gauge to reach its maximum reading. Then stop the engine and observe the vacuum gauge for ten or fifteen seconds.

If maximum vacuum is ten inches or more, and does not fall off when the engine is stopped, the flex line, pump valves, and filter bowl gaskets do not leak. If the maximum reading is less than 10 inches, or vacuum falls off rapidly, a leak is indicated.

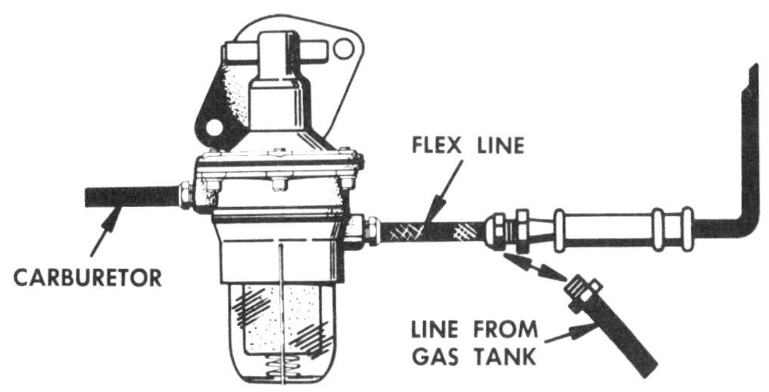

Fig. 10-13. Connection of vacuum gauge for fuel pump vacuum test.

The entire line from the fuel tank can be tested for air leaks in a similar manner by connecting the vacuum gauge to the fuel line at the tank end of the line, and repeating the test. A rapid fall-off of vacuum after the engine is stopped indicates air leaks into the line somewhere between the tank and the fuel pump.

Chapter 11

Exhaust Gas Analyzers

The most common exhaust gas analyzers in use are the thermal conductivity type and the infra-red type. The thermal conductivity exhaust gas analyzer was designed as a tool to set carburetors properly. With the advent of air pollution control, the infra-red type was developed in order to monitor exhaust emissions. This chapter will discuss both types of instruments.

THERMAL CONDUCTIVITY EXHAUST GAS ANALYZER

The thermal conductivity exhaust gas analyzer, Fig. 11-1, operates on the principle that the exhaust gas produced by different air-fuel ratios will have different cooling effects. For example, the exhaust gas produced by a rich air-fuel mixture will have a greater cooling effect than that produced by a lean mixture. This cooling effect is not dependent on the specific temperature of the exhaust gases but, rather, on their relative thermal conductivities. Thermal conductivity means the ability (or inability) of a substance to transfer heat.

Engine exhaust is composed of several gases, such as carbon dioxide, carbon monoxide, oxygen, hydrogen, nitrogen, and water vapor. The thermal conductivity of the exhaust gas varies according to the proportions of hydrogen ($H2$) and carbon dioxide ($CO2$) in the gas. The proportion of these gases in the exhaust varies and depends upon the air-fuel ratio of the mixture supplied by the carburetor.

A rich mixture, with more fuel and less air than normal, results in an increase in hydrogen and a decrease in carbon dioxide, and a lean mixture reverses the proportions. Hydrogen has a thermal conductivity about six times as great as air, while that of carbon dioxide is approximately one-half that of air. It is upon this difference in the thermal conductivity of the two gases and their varying proportions in the exhaust that the operation of the combustion analyzer is based.

For purposes of illustration, assume that air has a thermal conductivity of 1. By coincidence, exhaust gas produced from

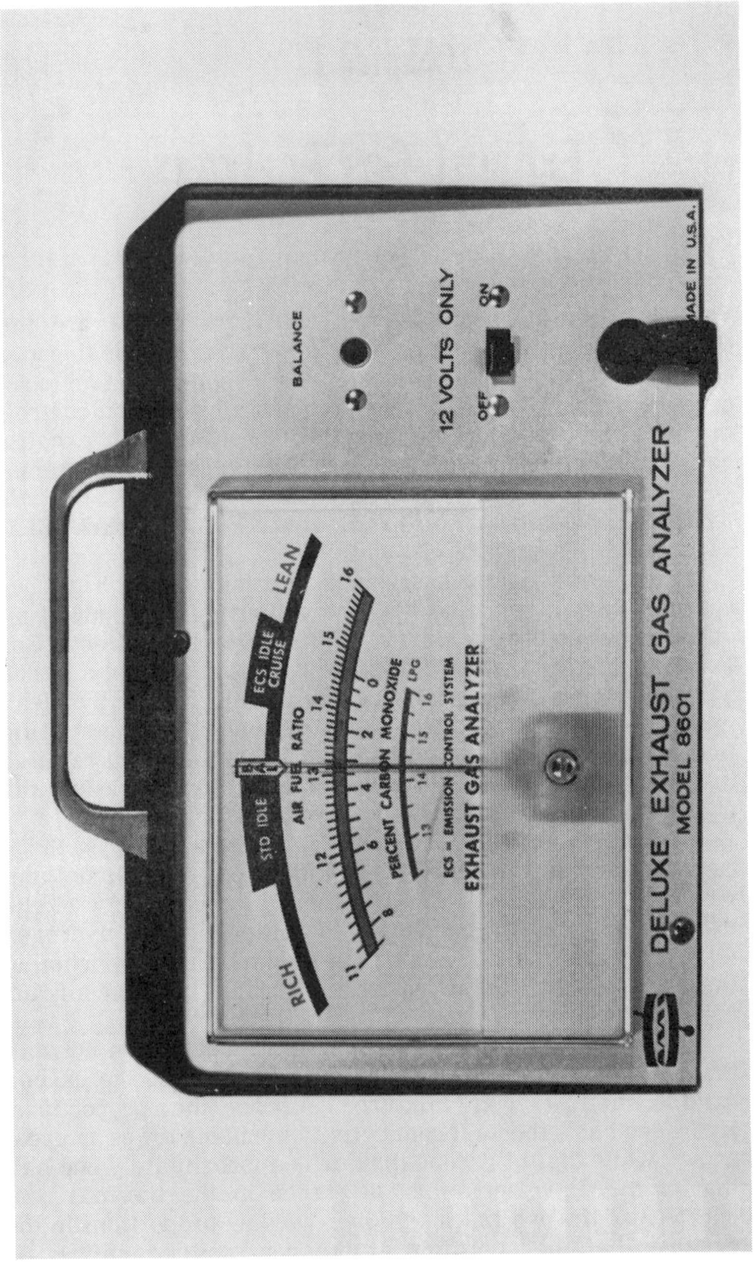

Fig. 11-1. Thermal conductivity exhaust gas analyzer.

an air-fuel ratio of 13.3 to 1 will also have the same thermal conductivity as air. Exhaust gas produced by an air-fuel mixture leaner than 13.3 to 1 will have a thermal conductivity of less than 1, or in other words, will have less cooling effect than air. On the other hand, exhaust gas from a mixture richer than 13.3 to 1 will have a thermal conductivity greater than 1, and therefore it will have a greater cooling effect than air.

This relative cooling effect of exhaust gas is put to use in the exhaust gas analyzer by passing it over a sensing unit which has been heated electrically to a certain temperature. The sensing unit can consist of either a platimum wire or a thermistor. The ability of the exhaust gas to cool this sensing unit (or allow it to become hotter) will depend on its thermal conductivity. As the temperature of the sensing unit varies, so will its resistance. The resistance of a platimum wire sensing unit will become greater as the temperature increases, while the resistance of a thermistor sensing unit will decrease as temperature increases. In either case, the change in resistance is measured on a meter, which is calibrated in terms of air-fuel ratio.

The sensing unit, over which the exhaust gas passes, is supported in a tube called an exhaust gas cell. Another sensing unit, identical to the one in the exhaust gas cell, is similarly supported in another tube called an air cell. The sensing units in these two cells form two branches of a Wheatstone bridge circuit, see Fig. 11-2. The main purpose of the air cell is to maintain a reference point while comparing the thermal conductivities of exhaust gas samples. No exhaust gas ever passes over the sensing unit in the air cell. It is, however, open to the air through a small hole in the side of the tube.

During the initial calibration of the tester, the sensing units in the exhaust gas cell and air cell are heated electrically to a certain temperature. This is done by passing a definite quanitity of current through the bridge, since the temperature of the sensing units depends on how much current is flowing through them. The fact that air has a thermal conductivity equal to that of exhaust gas from a 13.3 to 1 air-fuel mixture is used to establish a reference point for comparing other exhaust gas samples.

When exhaust gas from a rich mixture passes over the gas sensing unit, it will be cooled down to some lower temperature. Since the sensing unit is now at a lower temperature, its resistance will change, causing the potential at point B, Fig. 11-2, to change. The difference in potential between points B and A now cause the meter connected between these two points to read towards the rich end of the scale. If a lean

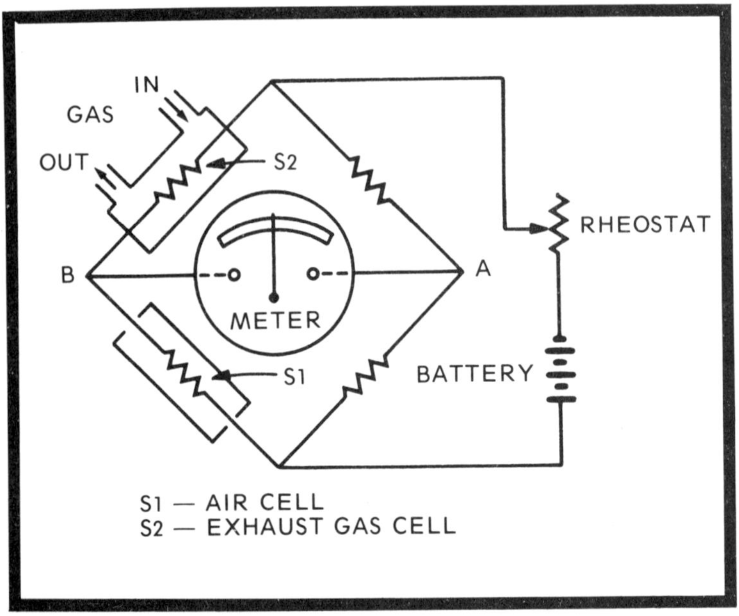

Fig. 11-2. Wheatstone bridge circuit for exhaust gas analyzer.

exhaust sample, leaner than 13.3, passes over the sensing unit, the sensing unit temperature and resistance will change; even more so than when only air was passing over the wire. This, again, changes the potential at point B, causing a difference in potential between points B and A. The meter will now read on the lean side of the scale.

Collecting Gas Sample

One of the requirements of all gas analyzers is to obtain a true sample of exhaust gas to analyze. This requires a gas sample pickup unit at the tail pipe, Fig. 11-3. The gas sample can then be directed into a sensing unit located at the tail pipe, or it can be sent through a hose to a sensing unit located in the instrument.

Instruments having a sensing unit located at the tail pipe send an electrical signal to the indicating instrument, which is usually located at the front of the car. The response time of this type of instrument is excellent in that it is very rapid. Response time is the interval of time it takes the tester to indicate a change in air-fuel ratio after a change has occurred in the engine.

The type of instrument in which the gas sample is sent through a hose to a sensing unit located within the instrument is usually provided with a booster pump. The booster pump reduces the response time of the instrument by drawing the gas through the hose to the sensing unit at a rapid rate.

Air dilution, that is, air being mixed with the exhaust gas sample, can change its thermal conductivity so as to cause inaccurate test results. This dilution can be prevented by taking the exhaust gas sample far enough up in the tail pipe to prevent air from being sucked into the pick-up unit. There is a continual, pulsating inrush of air between each exhaust pulse which can extend quite a distance up the tail pipe. The actual distance varies for different engines, but is usually greater for four cylinder engines than for six or eight cylinder engines. However, if the pick-up unit is long enough, air dilution from this cause can be prevented. Obviously, nothing can be done about air dilution caused by holes or cracks in the exhaust system except to repair them before making any combustion tests.

One enemy of proper operation is water condensation forming in the exhaust gas cell. This has a tendency to cause an erratic meter indication. The water produced during combustion often exceeds one gallon per gallon of fuel burned. Although this water is in the form of a vapor as it leaves the tail pipe, it tends to condense as it passes through the tester. To prevent this water from reaching the tester, it is purposely

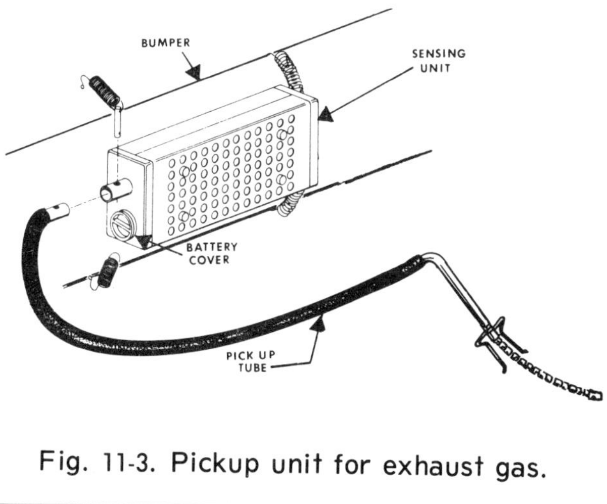

Fig. 11-3. Pickup unit for exhaust gas.

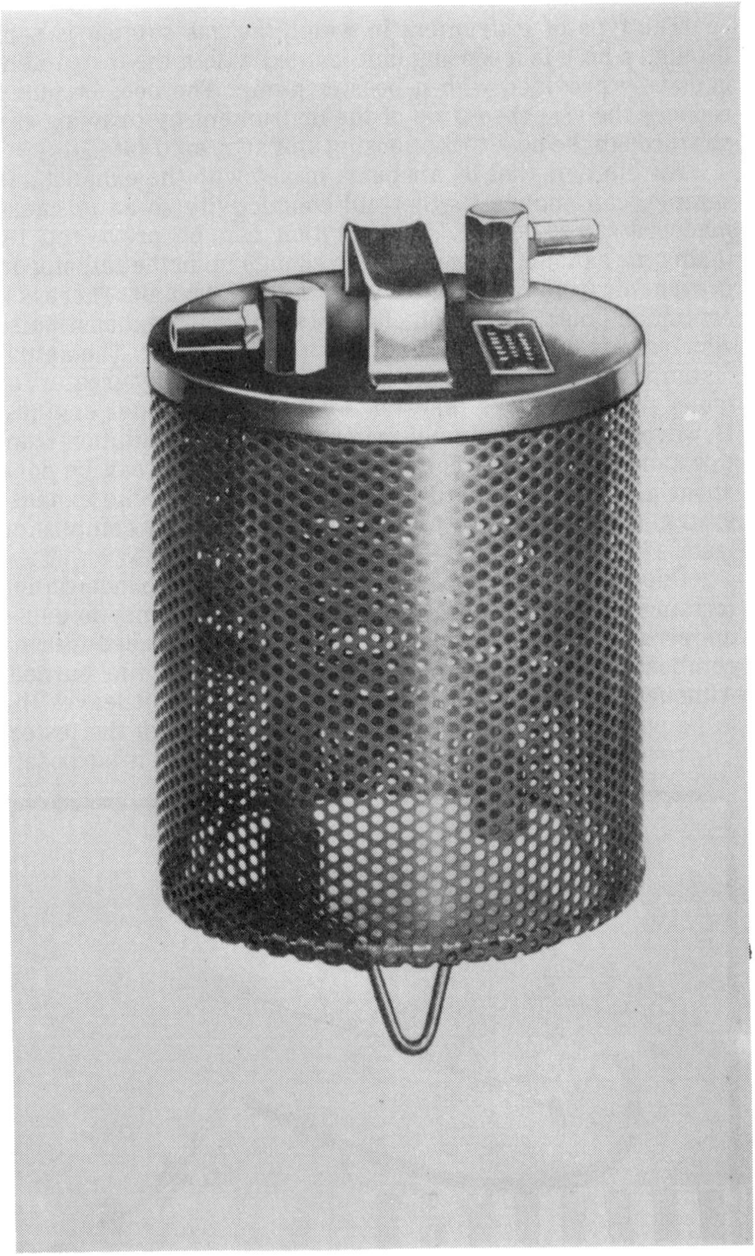

Fig. 11-4. Exhaust condenser.

condensed in an exhaust condenser, Fig. 11-4. Water traps which allow the water to drain from the condenser and at the same time prevent air from being drawn into the instrument are built into the lower part of the exhaust condenser. The exhaust condenser has the additional advantage of cooling the gas sample.

Never use the exhaust gas analyzer on an engine while gum solvents or oils are being introduced into the combustion chambers through the carburetor or by other means. Also, do not use the tester on an engine that is burning oil badly. Any oil leakage must be stopped before efficient combustion can be obtained, and the oil smoke will eventually decrease the sensitivity of the combustion meter.

Exhaust Gas Analyzer Meter

A typical exhaust gas analyzer meter dial is shown in Fig. 11-5. The dial shows two idle zones which will be explained later. The air-fuel ratio scale for an engine operated with gasoline is the top scale shown in the illustration. Two direct reading scales are provided for engines designed to operate on butane and propane gas. The scales are marked in air fuel ratio and correspond to the upper scale for gasoline engines except for their being staggered.

Other meter scales have a percent of combustion efficiency scale extending across the entire range of the meter and graduated from 60 to 100, see Fig. 11-6. This scale affords a

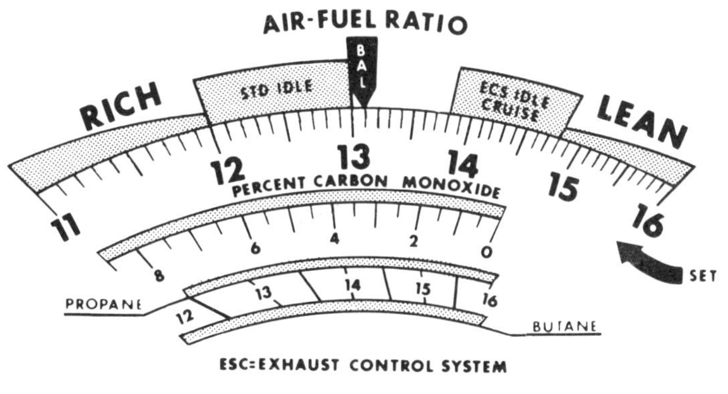

Fig. 11-5. Exhaust gas analyzer meter dial.

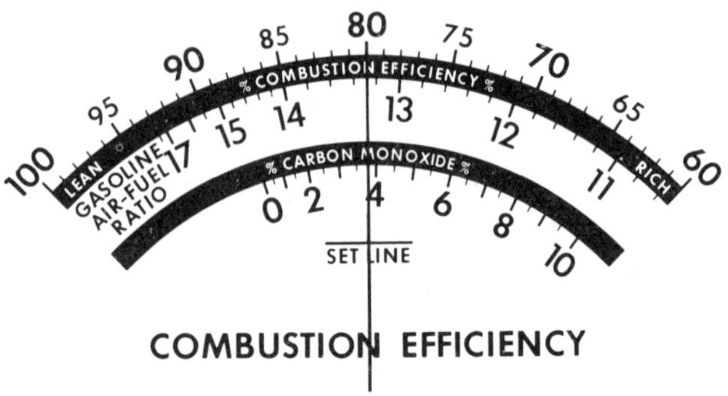

Fig. 11-6. Dial with combustion efficiency scale.

convenient reference when performing combustion tests. Through the use of this scale and adjoining chart, shown in Fig. 11-7, the test readings can easily be translated to air-fuel ratio values.

Since the percentage of carbon monoxide (CO) present in the exhaust gas of a motor vehicle is directly proportional to the air-fuel ratio, many exhaust gas analyzers include a Percent Carbon Monoxide scale on the meter dial as seen in both Figs. 11-5 and 11-6.

The thermal conductivity exhaust gas analyzer works well up to a certain point. This point, (called the stoichiometric point) is around 14.5 to 1 or a little higher. Gasoline is burned so completely at this time that the exhaust gas cannot further heat the thermal conductivity cell. Further leaning of the fuel mixture cools the cell and the meter starts reading in the wrong direction. Consequently, the value of the exhaust gas analyzer in practical tune-up and fuel system troubleshooting work is limited to showing the action of the carburetor within the power range of air-fuel ratios from 10.5 to 14.5

EXHAUST GAS ANALYZER TESTS

The exhaust gas analyzer can perform a double function. First, it can be used to obtain the best idle adjustment; and, second, it can be used to check the operation of the carburetor functions which come into play above idle speed. The adjustment that affects the air-fuel mixture is made on the idle circuit of the carburetor, but it also is important to know that

COMBUSTION EFFICIENCY Scale	AIR - FUEL RATIO Gasoline	Butane	Propane
60	10.50	10.90	11.10
61	10.60	11.00	11.20
62	10.75	11.15	11.30
63	10.90	11.30	11.45
64	11.00	11.40	11.60
65	11.15	11.55	11.75
66	11.30	11.70	11.90
67	11.45	11.80	12.00
68	11.60	11.95	12.15
69	11.75	12.10	12.30
70	11.85	12.20	12.45
71	12.00	12.35	12.60
72	12.15	12.50	12.75
73	12.30	12.65	12.90
74	12.45	12.75	13.05
75	12.60	12.90	13.20
76	12.75	13.05	13.30
77	12.90	13.20	13.45
78	13.00	13.35	13.60
79	13.15	13.50	13.75
80	13.30	13.60	13.85
81	13.45	13.75	14.00
82	13.60	13.90	14.20
83	13.70	14.00	14.35
84	13.85	14.20	14.55
85	14.00	14.40	14.75
86	14.20	14.60	14.90
87	14.45	14.90	15.10
88	14.65	15.15	16.00
89	14.90	15.40	17.00
90	15.50	16.00	-----
91	16.00	17.00	-----
92	16.60	-----	-----
93	17.30	-----	-----

Fig. 11-7. Table for transforming combustion efficiency scale readings to air-fuel ratios.

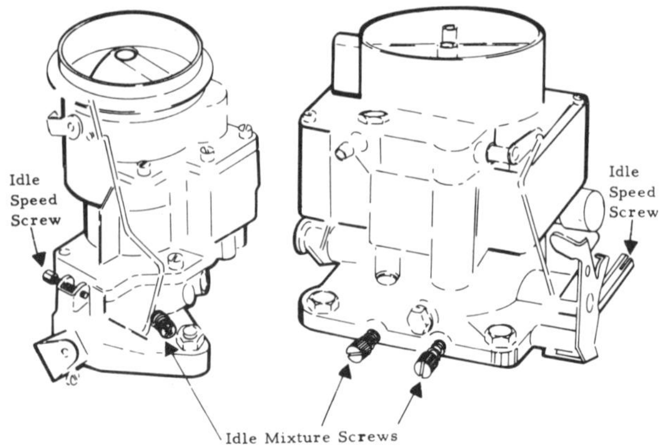

Fig. 11-8. Location of idle mixture screws.

the mixture is correct with throttle opening above idle and that the accelerating circuit is acting as it should.

The thermal conductivity exhaust gas analyzer tests that will be described are as follows:

1. Idle Adjustment
2. Accelerator Pump Test
3. High Speed Mixture Test

Idle Adjustment

After the engine has reached normal operation temperatures, shut the engine off. Turn the idle mixture screw(s) in until they bottom. Then back the screw(s) out three complete revolutions or until a stop is felt, whichever occurs first, see Fig. 11-8.

Start the engine and let it idle for one minute. If the car does not have an exhaust control system, turn the idle mixture screw(s) in ¼ turn. Wait until meter deflection is complete, and adjust idle speed to manufacturer's specification. Repeat this step until engine speed drops, or the engine starts to run rough. Then turn the screw(s) out ⅛ turn at a time until the engine speed is regained or the engine idle is smooth. The meter should indicate in the STD IDLE zone on the meter, see Fig. 11-9. If the reading is lean, turn the screw(s) out ⅛ turn more.

For vehicles equipped with an exhaust control system (ECS), refer to the decal in the engine compartment for idle adjustments. Disconnect and plug the manifold connection to the air injection pump, if the engine is so equipped. Where an air fuel ratio is specified, turn the idle mixture screw(s) in ⅛ turn. Wait until meter deflection is complete; and, then, using a suitable tachometer, adjust idle speed to manufacturer's specification. Repeat this procedure until the meter indication corresponds to the manufacturer's specifications. When completed, the meter indication should read in the ECS IDLE band shown in Fig. 11-9.

On multiple barrel carburetors a slightly finer adjustment can be obtained by turning each screw separately ⅛ turn either way to obtain maximum rpm as indicated by the tachometer.

If ECS IDLE or STD IDLE zones cannot be reached, check for and correct if necessary the following:

1. Dirty or plugged carburetor passages
2. Float level
3. Sticky needle valve
4. Automatic choke
5. Fuel pump pressure

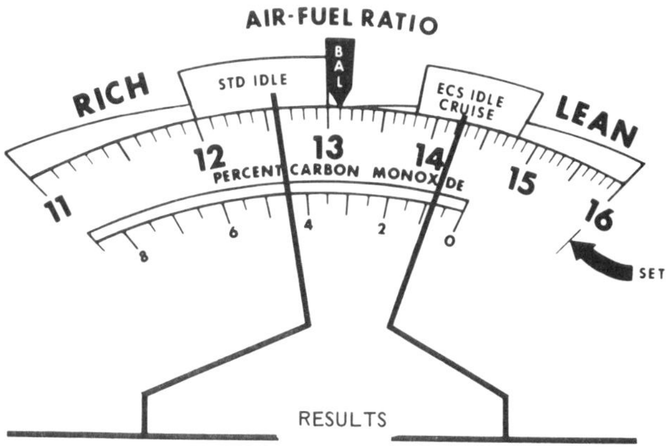

STD IDLE of cars without Exhaust Emission Control systems. Adjust for smoothest idle within STD. idle block.

ECS IDLE of cars with exhaust Emission Control systems. When adjusting these cars, it is important to follow the manufacturer's prescribed method because each car's system varies.

Fig. 11-9. Meter indications for idle adjustments.

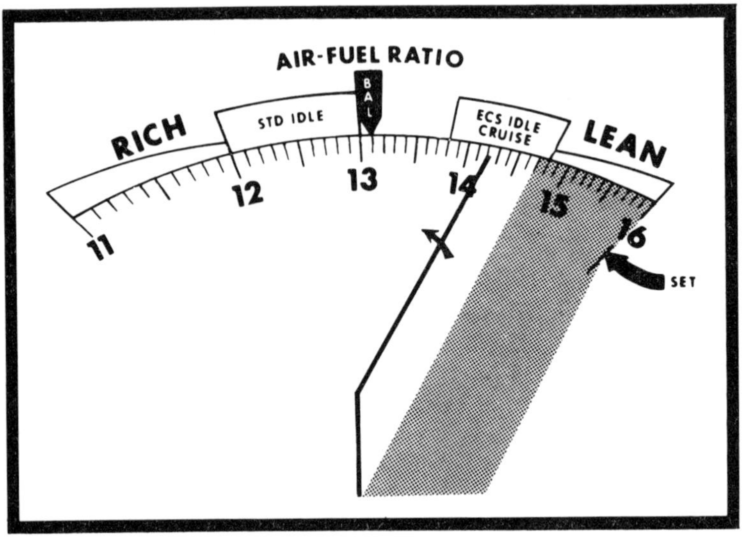

Fig. 11-10. Meter indications for accelerator pump test.

6. Ignition timing
7. Distributor advance, retard mechanism
8. Spark plug misfiring
9. Improper dwell
10. Clogged air filter
11. Carburetor flange and manifold leaks
12. Vacuum line leaks; distributor, windshield wiper, power units

Accelerator Pump Test

Set engine at 1000 rpm and note meter reading at that speed. Open the throttle wide and quickly return to 1000 rpm setting. Note meter reaction and the amount of movement toward the rich side of the scale. The pointer should move toward the rich side of the scale a few seconds after the throttle is opened and returned to its original 1000 rpm setting, see Fig. 11-10. Little or no change in meter reading indicates an insufficient pump stroke, leaky check valve, worn linkage, or worn plunger.

High Speed Mixture Test

This test is performed at an engine speed of 2500 rpm. The meter should read CRUISE or 13.8 or leaner at this speed, Fig.

11-11. A richer mixture indicates the following problems:

1. Clogged air filter
2. Choke stuck on
3. High float level
4. High fuel pump pressure
5. Main jets too large
6. Power circuit malfunction

An unsteady reading can be caused by:

1. Leaky or sticking needle and seat of carburetor
2. Manifold air leaks
3. Exhaust system leaks

An excessively lean reading can be due to the following:

1. Low float level
2. Weak fuel pump
3. Clogged gas filter

EMISSION CONTROL SYSTEMS

Reducing air pollution, or smog, is a matter of major importance since a high concentration of smog is not only

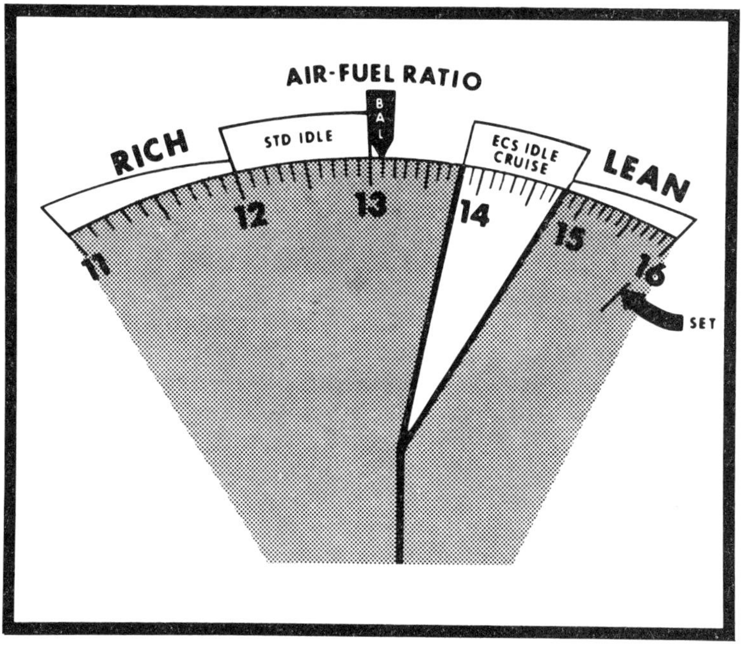

Fig. 11-11. Meter indication for high speed mixture test.

irritating and harmful to everyone, but can also seriously affect cattle and crops in agricultrual areas. Air pollution also results in less obvious but substantial losses because of deterioration of paint, metal, stone work, rubber, nylon and other textiles, and reduction of sunlight, in some cases by as much as 40 percent.

One major air pollution problem, that of photochemical smog, is a condition created by the reaction of sunlight with nitrogen oxides and hydrocarbons (unburned gasoline) which are released by automotive engine exhaust and crankcase vapors. Carbon monoxide, though it does not contribute to photochemical smog, is also a large contributor to air pollution.

To help remove these chemical emissions from motor vehicles, the automotive manufacturers have done extensive research in the field of emissions control. As a result, various emission controls have been added, and modifications have been made to recent automotive engines.

Crankcase emission (blowby vapors) previously vented to the air by way of the road draft tube is now controlled by a Positive Crankcase Ventilation system. Exhaust emissions (the product of combustion chamber burning of the air-fuel mixture), are reduced significantly by one of two systems: (1) an internal engine modification method or (2) an external after-burning air pump injector system.

The internal modification systems consist of specially calibrated carburetors, distributor, engine timing control vacuum lines, and the closed PCV system. The carburetors are designed for precise metering of fuel during the critical engine operating conditions created at idle. A very lean air-fuel ratio is obtained by this design.

The initial engine timing and timing advance curves have been changed to further reduce emission at idle and during deceleration. At speeds above idle, the engine timing returns to normal due to the modification of the distributor advance curve and the vacuum advance controls.

As to cooling system construction and capacity, emission control engines are designed to operate at higher temperatures to bring about more complete combustion within the engine. Thermostatically controlled air intake pre-heater systems are used in most vehicles to improve engine warm-up characteristics without the air of overly-rich (choke) mixtures.

The air pump injector system is generally provided for manual transmission vehicles and high performance engines. The air injection system utilizes low pressure air, injected into the exhaust ports near the exhaust valves, to further oxidize any unburned fuel which may escape into the exhaust system. In addition to the closed PCV system, this system consists of an air distribution manifold, a check valve, an anti-backfire or bypass valve, and the necessary interconnecting lines and tubes. Further information on how these various systems operate can be obtained from the manufacturer's operation and repair manuals.

In order to control the emissions from vehicles, maximum emission levels of hydrocarbons and carbon monoxide have been prescribed. Hydrocarbons (HC) are measured in parts per million (ppm) which means parts of hydrocarbon per million parts of air. Carbon monoxide (CO) is measured in percent. The instrument most commonly used to check the emission levels is the infra-red exhaust gas analyzer.

INFRA-RED EXHAUST GAS ANALYZER

The infra-red exhaust gas analyzer shown in Fig. 11-12 makes use of non-dispersive infra-red technology for measuring hydrocarbons and carbon monoxide. The operation of this unit is similar to the thermal conductivity type instrument already discussed—that is, a sample tube is inserted into the exhaust pipe, and after the instrument has been properly calibrated, a reading can be obtained from the meters.

Typical scales for this type of instrument are shown in Figs. 11-13 and 11-14. Carbon monoxide can be read from 0 to 7.5 percent, and hydrocarbons can be read from 0 to 2000 ppm (parts per million).

The value of this analyzer is shown in Fig. 11-15, which illustrates some of the diagnosis obtainable by taking a reading at idle rpm, 2500 rpm, and under rapid acceleration (goose test). For example, if the CO reading is low under all three conditions while the HC reading is high, a misfiring cylinder or cylinders should be suspected.

Carbon monoxide is a byproduct of combustion, and is not present in the exhuast until fuel is burned. Hydrocarbons, however, are components of the fuel and exist even when no combustion occurs, such as through simple evaporation of the gasoline or when a spark plug misfires. As combustion in a cylinder becomes more complete, the amounts of both carbon monoxide and hydrocarbons will decrease. Should a cylinder misfire, the lack of combustion will cause carbon monoxide to decrease. The misfiring cylinder now contains unburned fuel, which causes hydrocarbon readings to increase.

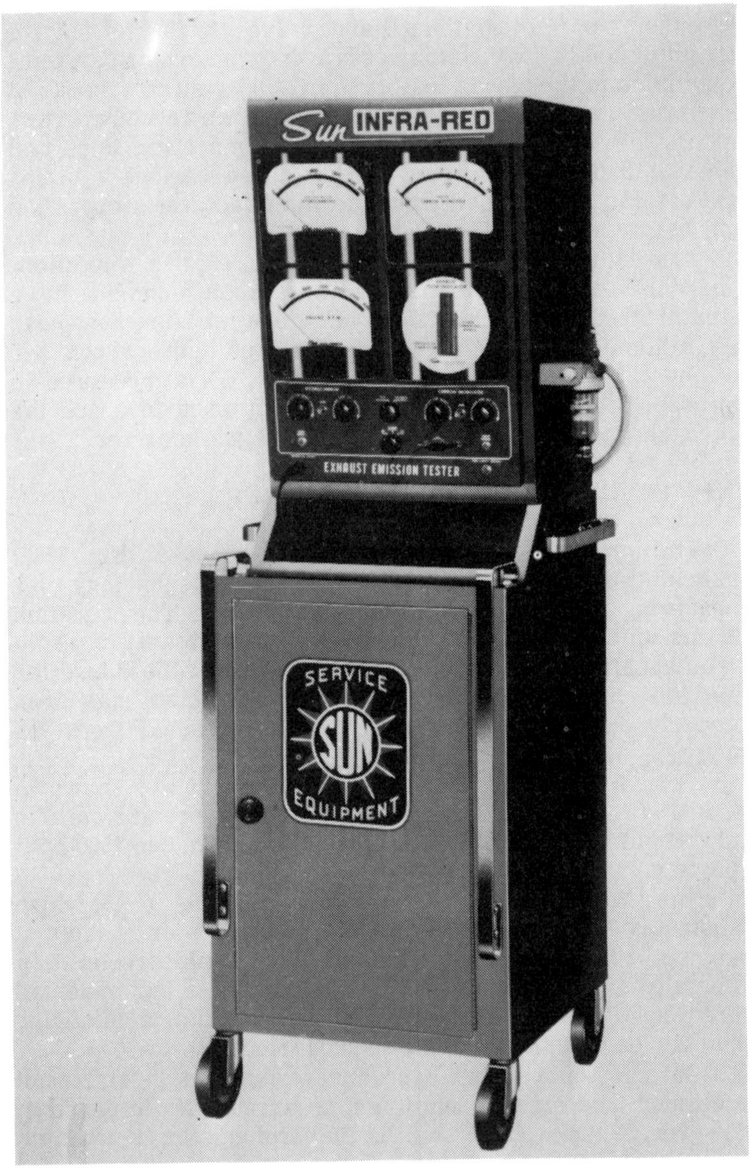

Fig. 11-12. Infra-red exhaust gas analyzer.

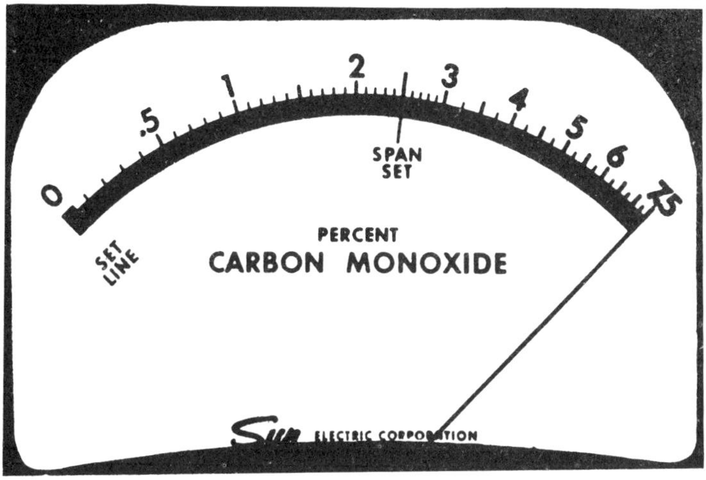

Fig. 11-13. Meter scale for carbon monoxide.

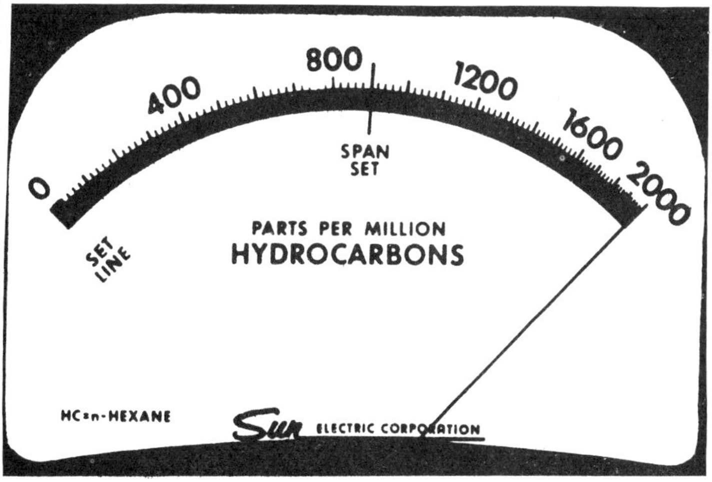

Fig. 11-14. Meter scale for hydrocarbons.

Operation of Infra-Red Exhaust Gas Analyzer

Gas containing hydrocarbons and carbon monoxide absorbs infra-red energy. If an infra-red beam is directed through an exhaust gas sample, some of the infra-red emergy is absorbed just as sunlight is absorbed by haze and fog. The infra-red energy that is not absorbed by the hydrocarbons and carbon monoxide is measured and electronically compared with the original amount of infra-red energy. The difference directly represents the amount of hydrocarbons or carbon monoxide present in the exhaust gas sample.

As shown in Fig. 11-16, exhaust gas is collected at the tail pipe (1) and drawn into the condenser (2) where it is cooled to remove excess moisture. The condensation flows into a water trap (3) and the gas continues through the particle filter (4) which removes solid particles. The gas then flows through the sampling tube (5). Infra-red energy, generated by a hot filament (6) is projected through the sample. The infra-red energy is chopped (7) and filtered (8) with an optical interference that passes a narrow band of infra-red at a

	TEST RESULTS			DIAGNOSIS
	Idle	2,500 rpm	Goose Test	
CO	Low	Low	High	Carburetor OK; Accelerator Pump OK
HC	Low	High	High	
CO	Low	High	High	Sticking choke; Dirty air cleaner; Defective carburetor
HC	Low	High	High	
CO	High	High	High	Sticking choke; Dirty air cleaner; High carburetor float level
HC	High	High	High	
CO	Low	Low	Low	Misfiring cylinder or cylinders
HC	High	High	High	

Fig. 11-15. Test results and diagnosis for various engine conditions.

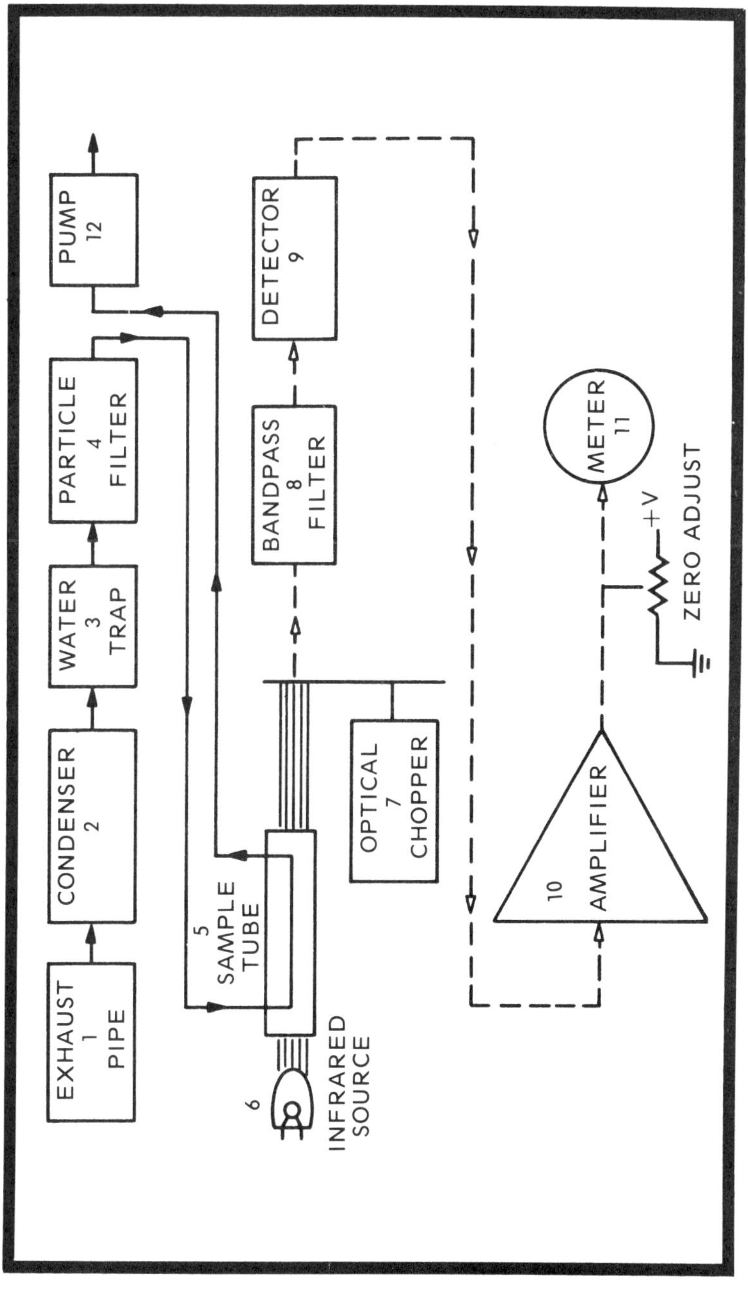

Fig. 11-16. Block diagram of the operation of the infra-red exhaust gas analyzer.

wavelength at which either carbon monoxide or hydrocarbons has the strongest absorption band. The infra-red energy is focused on the detector (9) which converts the infra-red energy into an electrical signal. The signal is then passed through an amplifier (10) to the meter (11). The gases are physically drawn through the analyzer by a vacuum pump (12).

In an instrument that measures both carbon monoxide and the amount of hydrocarbons the procedure is repeated: two infra-red sensors are used, one adjusted to read carbon monoxide and the other adjusted to read hydrocarbons. Due to the non-dispensive characteristics of the instrument, the same gas sample can be passed on from the carbon monoxide detector to the hydrocarbon detector.

Chapter 12

Specialized Testers and Diagnostic Equipment

Most of the instruments previously discussed are used for general tune-up work or system analysis. There is also equipment for testing specific components and equipment for overall trouble diagnosis.

SPECIALIZED TESTERS

Two useful instruments for performing specific tests are the coil-condenser tester and the distributor tester.

Coil Tester

An ignition coil consists of two windings: a primary winding of a few hundred turns of relatively heavy wire and a secondary winding of many thousands of turns of very fine wire. Both are wound on the same core, with the primary winding customarily wound on the outside of the secondary winding. The primary and secondary windings are insulated from each other.

The relatively few turns of the fairly heavy primary winding has a low resistance which will permit a large current flow when the breaker points are closed, resulting in the build-up of a magnetic field. When the points open, the current stops flowing and the magnetic field collapses. As it collapses, the magnetic lines of force cut across the conductors of the secondary or high tension winding and induce a high voltage sufficient to break down the rotor and spark plug gaps.

The coil testing portion of the coil-condenser tester shown in Fig. 12-1 operates by stressing a coil to 20 kilovolts, at which point the oscillations displayed on a scope are observed. The tester electronically allows a magnetic field to build-up and collapse, and this action is controlled by a knob on the front panel. The pattern on the screen can be checked for shorts, opens, grounds, and insulation breakdown within the coil.

If the waveform height is 20 KV or more, and the oscillations are similar to those shown in Fig. 12-2, the ignition

Fig. 12-1. Example of coil-condenser tester.

coil is in satisfactory condition. When few oscillations are seen, the coil is defective due to a shorted primary or secondary winding or internal insulation breakdown, Fig. 12-3. An unstable or jittery pattern also indicates a defective condition due to internal insulation breakdown.

When nothing more than a straight line is seen on the scope, the coil has an open primary winding.

Condenser Tester

The purpose of the condenser in the ignition system is to prevent arcing and pitting of the breaker points, and to aid in

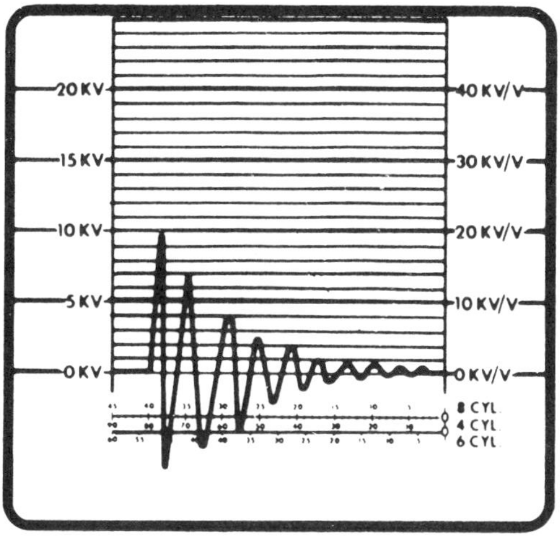

NORMAL

Fig. 12-2. Normal pattern for good coil.

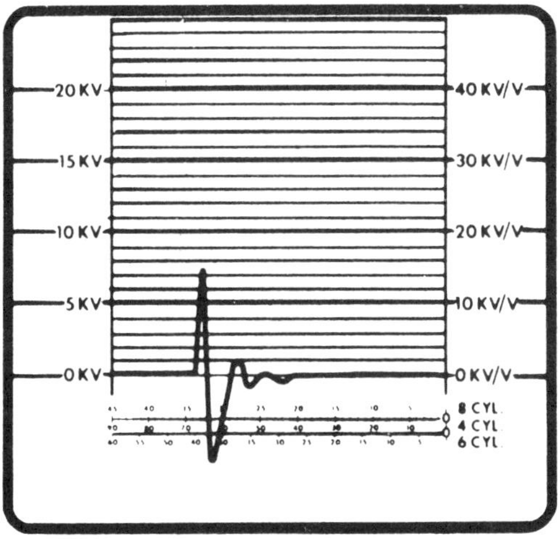

SHORTED

Fig. 12-3. Pattern for defective coil.

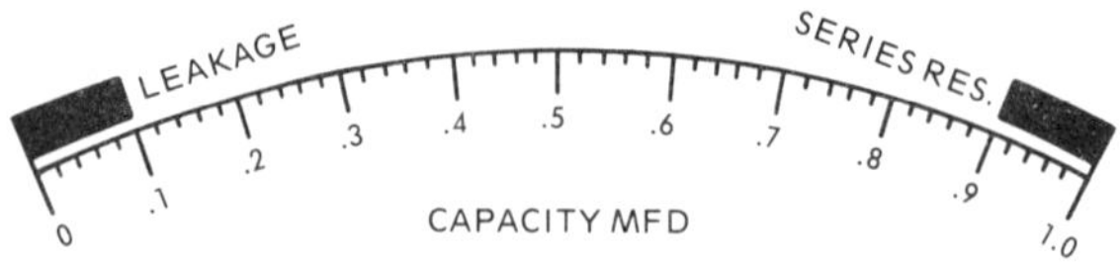

Fig. 12-4. Condenser meter scale.

collapsing the magnetic field in the ignition coil. In order to function properly and assure good ignition, the condenser must have three important characteristics, which are:

1. Minimum series resistance to current flow into and out of the condenser:

2. Correct capacity for the ignition system in which it is used:

3. Minimum insulation leakage so it will hold the electrical charge as it should.

Each of these factors is tested by the condenser testing portion of the coil-condenser tester shown in Fig. 12-1, and the results of the three tests indicate whether the condenser is usable or should be replaced.

To test the series resistance of a condenser requires a circuit quite different from the ohmmeter used to make conventional resistance measurements. This is because the condenser does not permit the flow of direct current through it, and most ohmmeters are DC instruments.

One way of measuring the condenser series resistance is to make use of a circuit which is basically a high frequency oscillator with a meter to measure the strength of the oscillations. The condenser to be tested must be connected into the oscillator circuit by means of the test leads. When this is done, the circuit will oscillate. How well it will oscillate depends on the series resistance of the condenser connected into the circuit. If the series resistance is low, as it should be, the circuit will oscillate at or near full strength and the meter will show almost full scale deflection. This amount of deflection is usually indicated on the meter by a colored bar, within which the needle should point if the condenser is usable, see Fig. 12-4.

Condenser capacity is measured by a circuit in which an intermittent current flows to charge and discharge the condenser under test. The condenser capacity is proportional to the discharge from the condenser which in turn affects the reading of the meter. The larger the capacity of the condenser, the greater the electrical charge it will store, resulting in greater meter deflection as it discharges through the meter. Since the charging and discharging take place very rapidly the meter will indicate a steady value.

The unit of condenser capacity is the microfarad, abbreviated as mfd. The range of a good ignition condenser will ordinarily be between .15 mfd and .30 mfd.

The condenser operating in an average ignition system is subjected to an electrical stress of 200 to 300 volts. Unless the condenser can withstand at least this voltage, the insulation between the plates of the condenser will soon break down, short the condenser, and stop the engine. The insulation leakage circuit of the tester simulates the stress of the ignition system by applying approximately 500 volts DC to the condenser during the test. This circuit is essentially an ohmmeter which measures the amount of leakage resistance between the condenser plates.

If a condenser with no leakage is connected to the test leads, no current will flow through the meter, and the meter will indicate infinite resistance. If the condenser is leaky, some current will flow through it and also through the meter. The amount of leakage resistance can then be read on the meter. To simplify reading the meter, the allowable amount of leakage resistance is indicated by means of a colored bar on the meter scale, see Fig. 12-4.

Ignition condensers may be tested in or out of the vehicle. If they are to be tested in the vehicle, it is necessary to block open the breaker points and to disconnect the wire from the condenser.

Distributor Testers

Distributor testers are variable-speed devices into which a distributor is clamped for the purpose of checking the mechanical and vacuum advance mechanisms. Many distributor testers come equipped with tach-dwell meters, a vacuum gauge, and a condenser tester, see Fig. 12-5.

The distributor is connected to a disk which revolves at the same speed as the distributor shaft. A light similar to a timing light flashes each time the distributor breaker points open, illuminating an arrow on the rotating disk. The effect, as the arrow rotates, is the appearance of eight arrows equally spaced around the circumference of a degree ring for an eight-lobe distributor, or six arrows for a six-lobe distributor, see Fig. 12-6. At a distributor speed corresponding to engine idle speed, with no vacuum, the degree ring, which can be moved

Fig. 12-5. Example of typical distributor tester.

around the center of the distributor shaft drive by the operator, is adjusted so that one of the arrows points to 0. The other arrows then should be spaced 45 degrees apart for an eight-lobe distributor or 60 degrees apart for a six-lobe distributor.

The operator should carefully observe each of the arrows to see that each one is at its corresponding mark on the degree ring. It is possible that one or more of the arrows may vary from 1 to 4 degrees from its correct mark, see Fig. 12-7. If this happens, it indicates that the distributor breaker cam is unevenly worn, the points are bouncing, or there is wear in the distributor shaft bushings which causes variations in the firing time of the individual cylinders. When this condition is observed, the distributor should be rebuilt or replaced.

The mechanical advance mechanism of the distributor is tested by increasing the speed of the distributor and noting how many degrees the arrow shifts from its initial position. In Fig. 12-8 the initial position is represented by the solid arrow. With an increase in speed, the arrow will shift to a new position represented by the dashed arrow. The amount of advance in degrees is the number of degrees difference between the two arrows as measured on the degree ring.

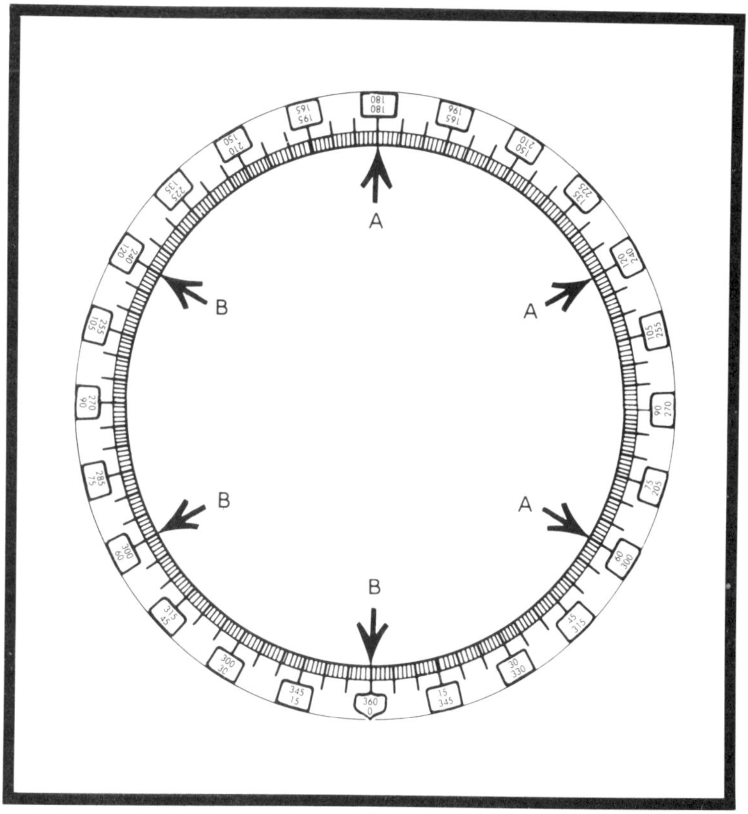

Fig. 12-6. Illuminated arrows on distributor tester for a six-cylinder engine.

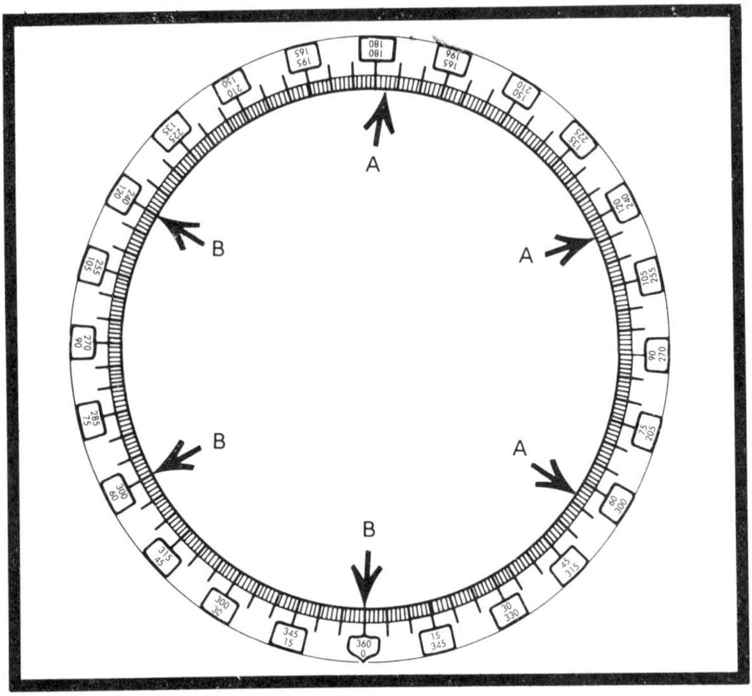

Fig. 12-7. The arrows marked A are off 3 degrees, while those marked B are correct.

If the advance is excessive, the governor weight springs are weak, or the wrong springs are installed. On the other hand, if the advance is insufficient both on acceleration and deceleration, the governor spring tension is excessive.

The vacuum advance mechanism is tested with a built-in source of vacuum in the distributor tester. The proper amount of vacuum is applied to the distributor for each of the specified check points and the amount of advance is obtained from the degree ring in a similar manner as illustrated in Fig. 12-8.

Excessive advance indicates a weak vacuum advance spring while insufficient advance indicates an excessively strong advance spring.

DIAGNOSTIC EQUIPMENT

Combinations of instruments are available as automotive analyzers, tune-up kits, and consoles. All of the testing features of these instruments have already been explained in the chapters dealing with the individual testers.

Automotive Analyzers

Automotive analyzers combine different testers into one package. The obvious advantages of an automotive analyzer are cost savings and compactness. For example, Fig. 12-9 illustrates a typical automotive analyzer with the following test features:

Voltmeter
Ammeter
Point resistance test
Dwell
Low and high rpm
Spark output test
Ohmmeter

All of this testing ability is contained in a package measuring 9" x 5½" x 3."

A somewhat more elaborate automotive analyzer is shown in Fig. 12-10. This tester has the following functions:

Three voltmeter ranges
High and low ohmmeter scales
Point Resistance test
Dwell
Low and high rpm
Alternator test
Ammeter
Condenser
Spark output test

When the automotive analyzer is used in conjunction with a timing light and vacuum and compression gauges, a substantial number of tests can be performed on an engine. With these few instruments, a skilled operator can turn out a professional tune-up.

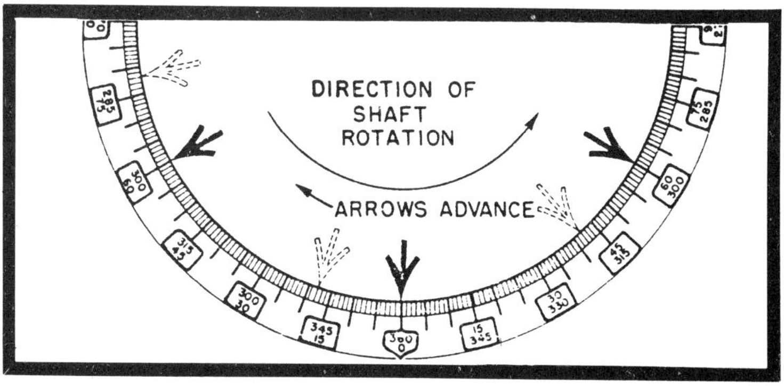

Fig. 12-8. Indications for obtaining spark-advance measurements.

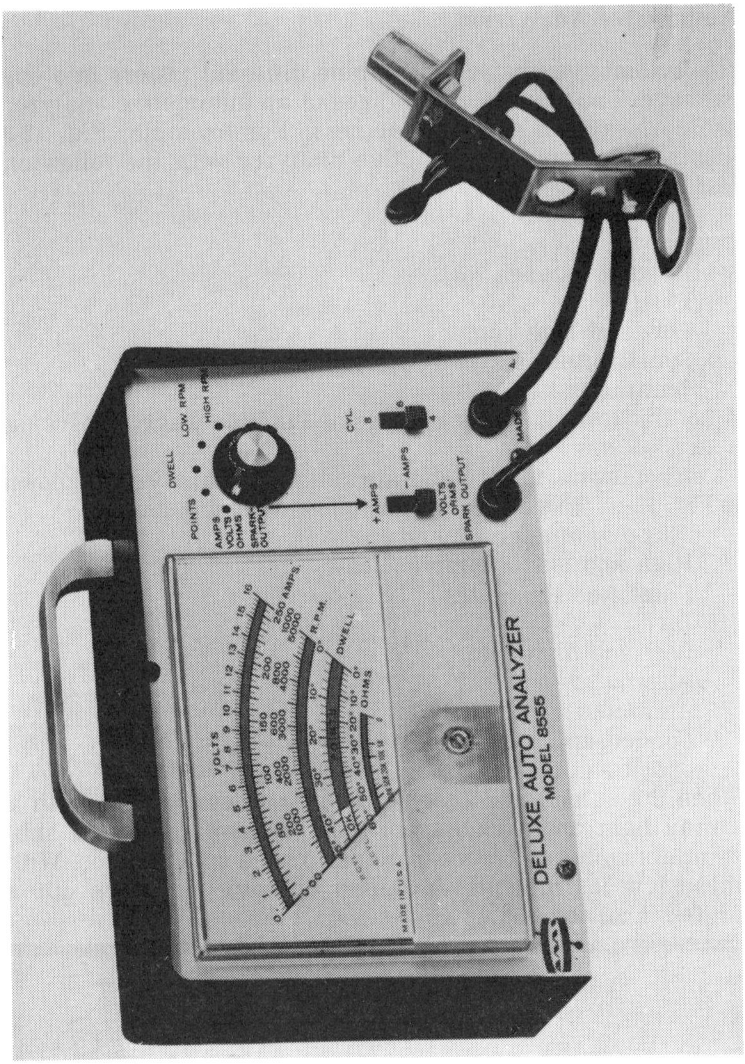

Fig. 12-9. Typical example of simple automotive analyzer.

Tune-Up Kits

For the person who would rather make one purchase of test equipment and get everything he needs at one time, the tune-up kit is the answer. A tune-up kit frequently contains three or four separate instruments with a comprehensive instruction procedure expalining how to perform a tune-up with the testers. Fig. 12-11 shows a tune-up kit containing a Tach-dwell, alternator-generator-regulator tester, and an ignition tester.

Another tune-up kit is shown in Fig. 12-12. This kit contains a tach-dwell, timing light, remote starter switch, vacuum gauge, and compression tester.

An individual who is just beginning to become familiar with automotive testing will usually find it to his advantage to buy a low cost kit in order to become better acquainted with the instruments. It will also serve as a base of experience for the selection of instruments in the future.

Consoles

Whereas a tune-up kit consists of several separate pieces of equipment, a console tester is an integration of many in-

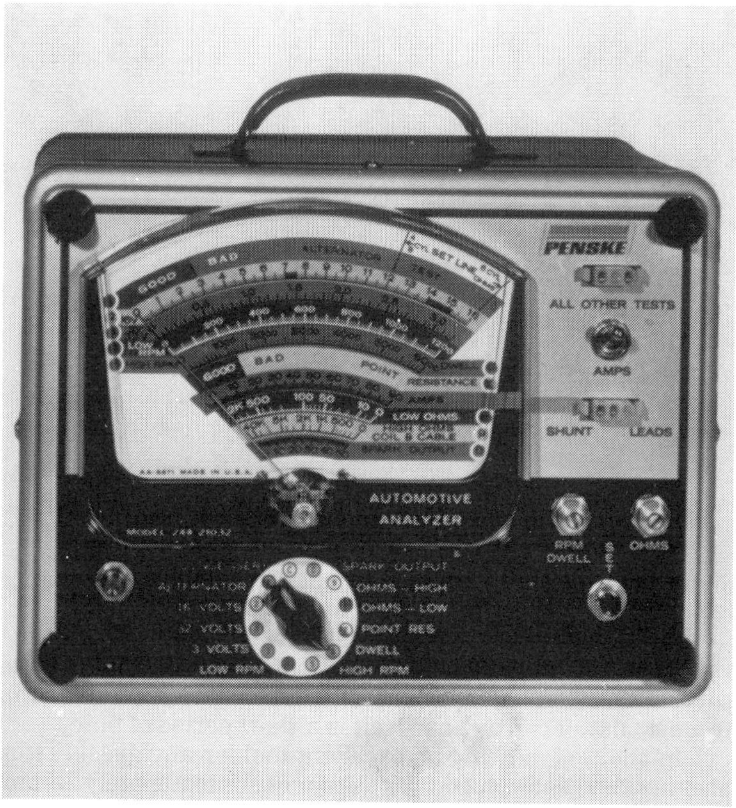

Fig. 12-10. Example of automotive analyzer.

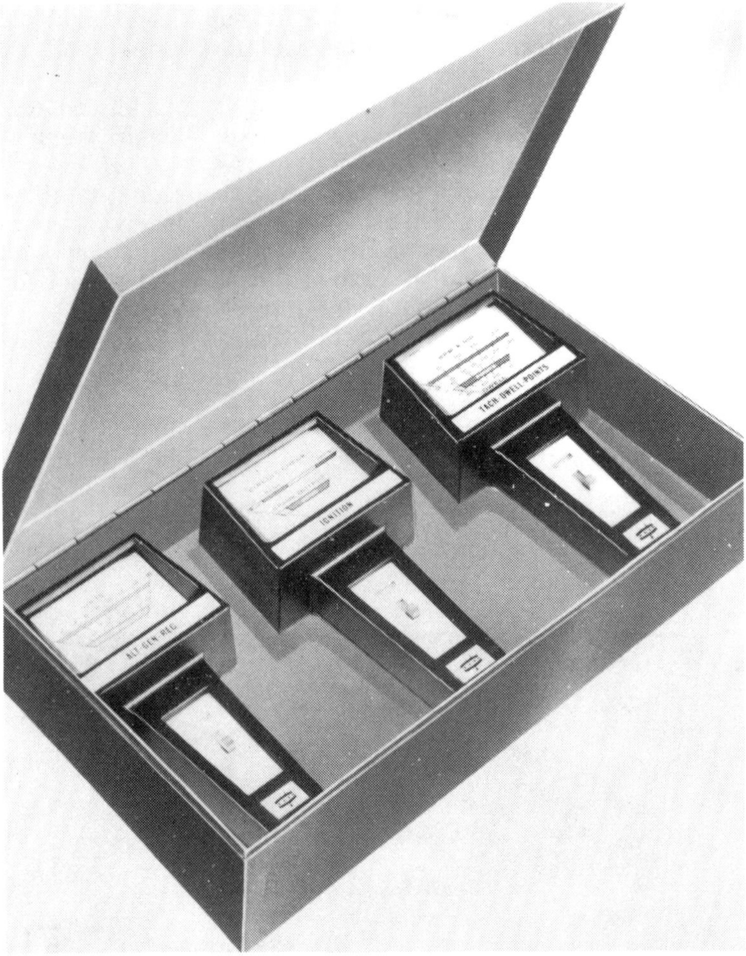

Fig. 12-11. Typical example of simple tune-up kit.

dividual testers into one instrument. The general appearance of a console is as one huge tester, see Fig. 12-13.

The use of a console is generally restricted to the full-time skilled mechanic because the cost of a console can run to several thousand dollars. Nevertheless, where a console is in frequent use, it can pay for itself in a short period of time.

In addition to being an excellent tool for any garage, it is also an ideal showpiece for customers. Unfortunately, many consoles sit unused in garages where the mechanics have not taken the time to learn how to operate them. If you purchase a

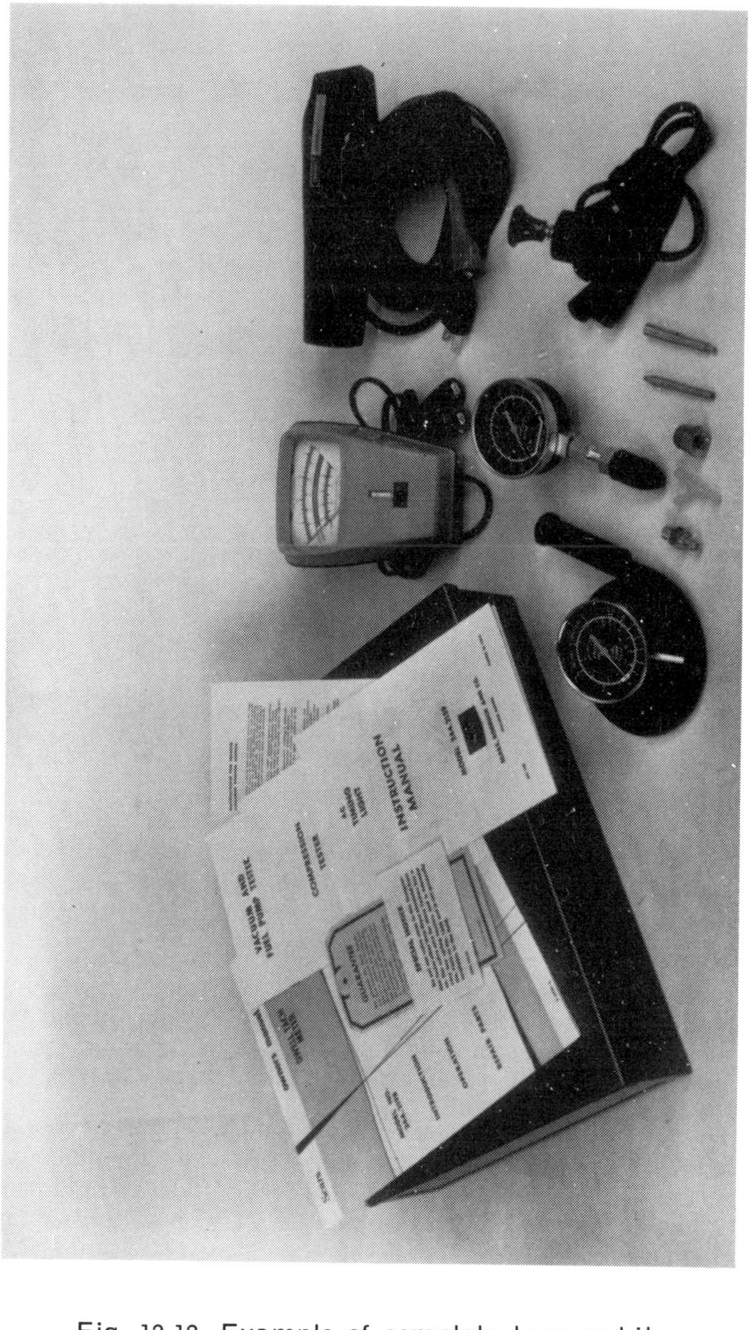

Fig. 12-12. Example of complete tune-up kit.

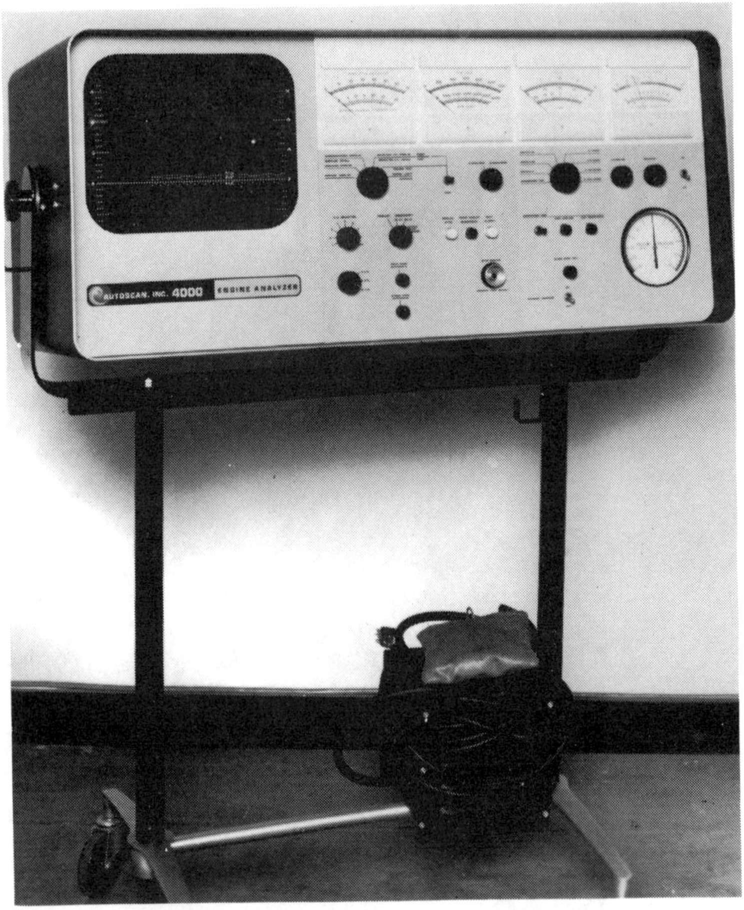

Fig. 12-13. Typical example of console.

console, learn how to use it, so that the confidence a customer has in you because of the appearance of a console will be justified. It may seem incredible that anyone would spend thousands of dollars on an instrument and then not use it, but this happens all too often.

Actually, learning how to use a console for tune-up work is surprisingly simple, if you know the basics of the individual testers discussed in the previous chapters. Always remember that a console is only a combination of many simplier instruments.

A very distinct advantage of a console over an assortment of individual testers is that the lead hook-up can be simplified. If one were to hook-up all the individual testers that comprise a particular console, he would find that there are many common leads. For example, a connection to ground is required for most instruments, such as the tach-dwell, voltmeter, timing light, and ignition tester. Also, a connection to the distributor side of the coil is required for the tach-dwell tester, scope primary pattern, cylinder shorting unit, and some ignition testers. In addition, a trigger pick-up is needed for both the scope and cylinder shorting unit. A console combines the common connections into one lead so that instead of having, for example, numerous ground leads, one wire is all that is necessary.

Most consoles have five or six leads to connect, as shown in Fig. 12-14. The lead connections are made to the same places on the engine as with the individual testers. The leads are usually combined into a harness for ease of handling. One thick cable comes out of the console which then splits into the individual leads near the end.

Another significant advantage of a console is the speed with which an engine can be tested. Most consoles are designed around a logical step-by-step procedure. The procedure automatically guides the mechanic through all of the tests the console can perform. The test results are noted on a check-off list which can be given to the customer along with recommendations for repairs or adjustments.

If a customer gives the go ahead to make a repair, the results of the work can be checked by retesting with the console. In using a console in this manner it acquires two separate functions. First, it can be used to make a quick overall check on an engine to determine its condition. Next, it can be used to verify that the proper repair or adjustment has been performed.

Several variations in procedures for consoles are in common use. One procedure is to perform the tests as a function of speed. That is, several tests are performed while the engine is cranking. Several more tests are performed at idle speed, and a final group of tests are performed at 1000 rpm and 2500 rpm. A check-off sheet for this procedure is shown in Fig. 12-15.

Another common procedure is to make all the tests that pertain to a specific system in the engine one at a time. For example, first all the tests on the starting system are made. Next, the charging system is tested, followed by the ignition and duel system tests.

Yet another procedure follows the layout of the front panel of the Console. Fig. 12-16 shows an example of a scope-less console that follows this procedure. The tests follow one another in sequence across the face of the console.

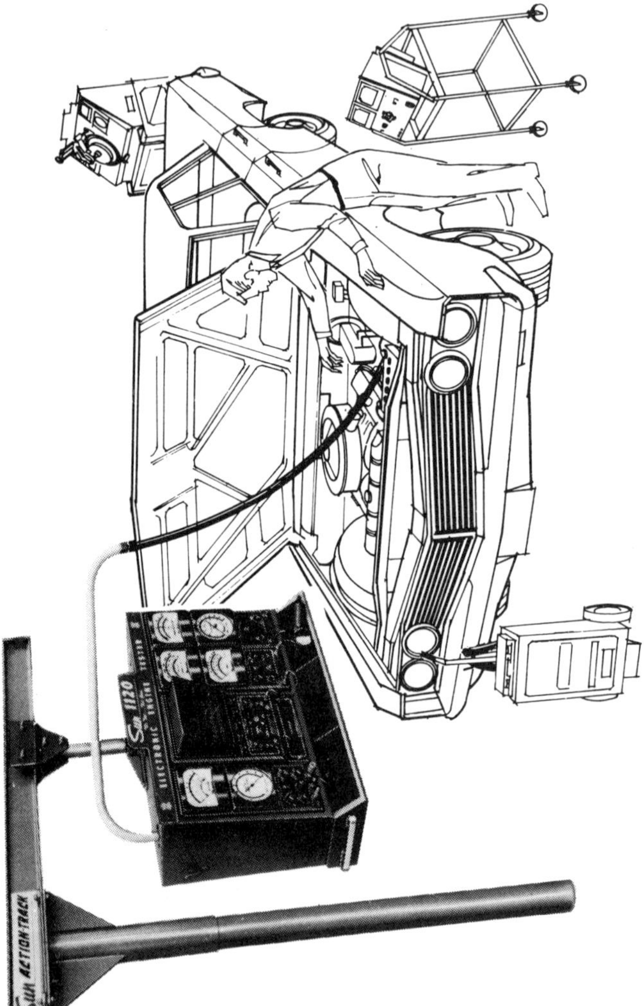

Fig. 12-14. Test leads made into a cable on console.

Consoles are constructed in two different ways. One type consists of one front panel with all the meters and internal circuitry mounted on that panel. The other type consists of several separate panels mounted on a stand and interconnected.

The advantage of the single panel instrument is that the controls for the instrument can be arranged on the panel so as to simplify the appearance of the console and its use. On the other hand, the advantage of the multi-unit console is that in the event of a failure, only the defective portion of the console need be sent back for repair, while the remainder can still be used to perform tests.

A building-block approach is used with some multi-unit consoles. Fig. 12-17 shows a console in which five separate and

CUSTOMER'S COMMENTS:

	TEST PROCEDURE	READ	SPECS.	RESULTS	GOOD	BAD
START	Cranking Voltage	Voltmeter				
	Cranking Coil Output	Scope (Display)				
	Cranking Vacuum	Vacuum Gauge				
	PCV Test	Vacuum Gauge				
IDLE	Idle Speed	Tachometer				
	Dwell	Dwell-Meter				
	Initial Timing	Timing Advance Unit				
	Fuel Mixture	Combustion Eff. Tester				
	Manifold Vacuum	Vacuum Gauge				
CRUISE	Dwell Variation	Dwell Meter				
	Coil Polarity	Scope (Display)				
	Spark Plug Firing Voltage	Scope (Display)				
	Maximum Coil Output	Scope (Display)				
	Secondary Circuit Insulation	Scope (Display)				
	Secondary Circuit Condition	Scope (Raster)				
	Coil and Condenser Condition	Scope (Raster)				
	Breaker Point Condition	Scope (Raster)				
	Cam Lobe Accuracy	Scope (Superimposed)				
	Fuel Mixture	Combustion Eff. Tester				
	Cylinder Power Balance	Tachometer				
	Record RPM	1 2 3	4 5	6	7	8
ACCELERATION	Spark Plugs Under Load	Scope (Display)				
	Accelerator Pump Action	Combustion Eff. Tester				
HIGH SPEED	Timing Advance	Timing Advance Unit				
	Charging Voltage	Voltmeter				
	Fuel Mixture	Combustion Eff. Tester				
	Exhaust Restriction	Vacuum Gauge				

Fig. 12-15. Check-off sheet for diagnostic procedure.

distinct instruments are combined. Each instrument can be purchased separately, or just the bottom three units can be built into a console, or all five units can be combined as shown. In addition, all of the units are quickly removable from the console and may be used separately whenever an occasion arises.

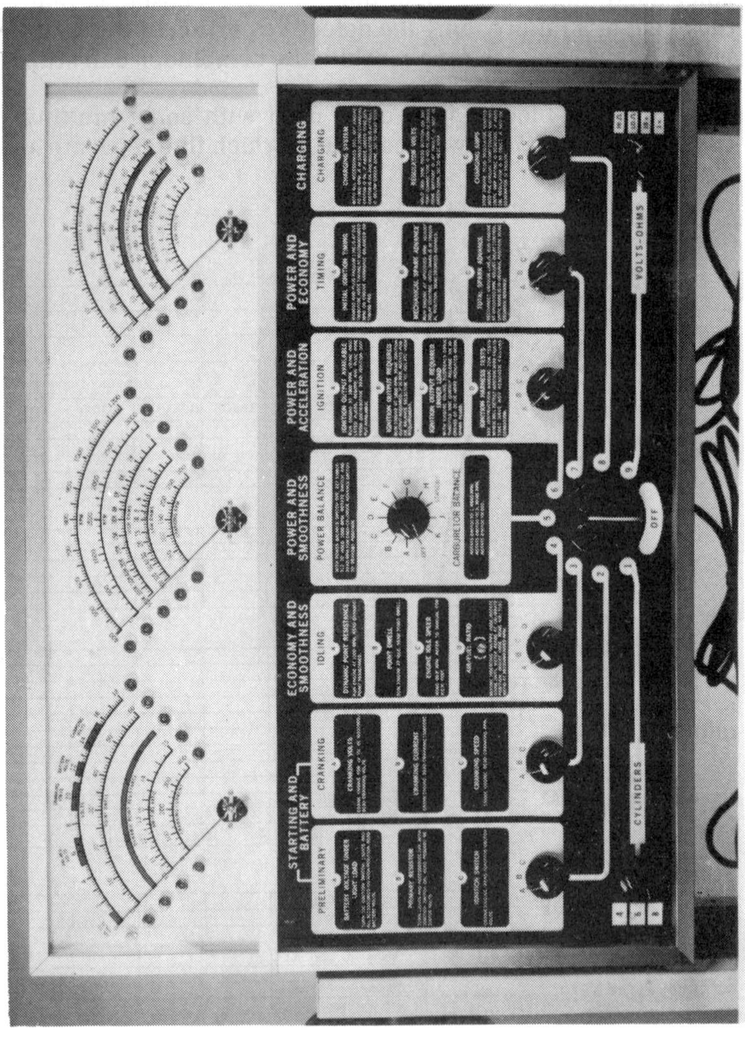

Fig. 12-16. Procedure based on front panel layout.

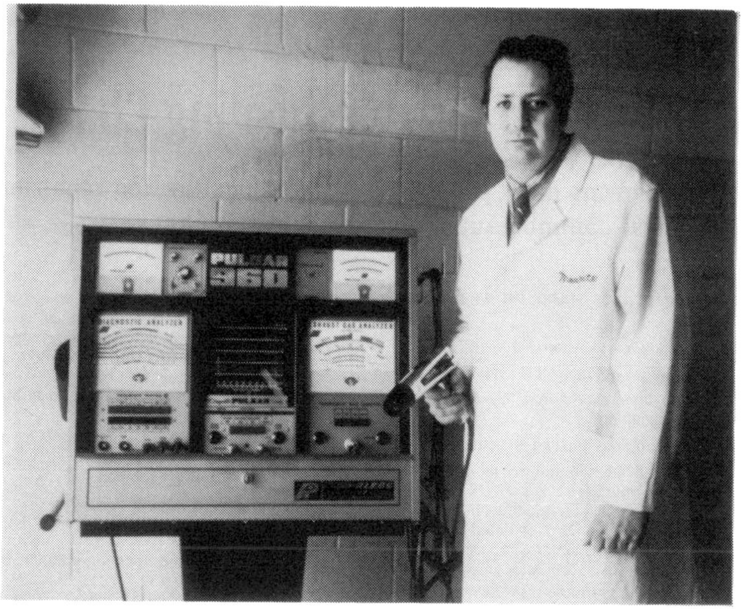

Fig. 12-17. Console built up from individual units.

Appendix 1
Tune-up Troubleshooting Chart

The following chart provides a guide to locating and repairing the most common automobile trouble symptoms.

ENGINE STARTS HARD OR WILL NOT START

1. Improper carburetion due to:

- A. Choke not operating properly.
- B. Cylinders and manifold flooded with gasoline.
- C. Carburetor passages restricted.
- D. Insufficient quantity or lack of gasoline in carburetor.
  1. Partially clogged fuel filter.
  2. Clogged or restricted fuel lines from gas tank to carburetor.
  3. Leaking carburetor gaskets.
  4. Low carburetor fuel level.
- E. Defective fuel pump.
- F. Air leaks at intake manifold or carburetor due to:
  1. Loose manifold nuts or cap screws.
  2. Leaking intake manifold or carburetor gaskets.
  3. Leaks occurring in vacuum line connections at intake manifold.
  4. Leaks occurring from vacuum operated accessories.
  5. Warped carburetor or manifold attaching flanges.
  6. Cracked intake manifold.
- G. Poor grade or old fuel combined with cold weather.
- H. Vent in fuel tank filler cap clogged or restricted.

2. Electrical difficulties.

- A. Battery
  1. Low or completely discharged.
  2. Terminals loose or badly corroded.
  3. Improper ground.
  4. Cables frayed or undersize.
  5. Battery capacity too small for the job.
  6. Charging circuit not operating satisfactorily.
- B. Ignition
  1. Primary circuit.
     - a. Broken wire lead between coil and distributor.
     - b. High circuit resistance due to corroded, dirty, or loose connections.
     - c. Ignition out of time.
     - d. Weak or grounded condenser.
     - e. Distributor breaker points improperly spaced, dirty, or loose.
     - f. Breaker arm sticking, spring weak or broken, or arm grounded.
     - g. Loose or grounded distributor terminal post.
     - h. Open ballast resistor (12 volt system).
     - i. Defective ignition coil.
  2. Secondary circuit
     - a. High resistance or open T.V.R. cables.
     - b. Corroded secondary cable terminals.
     - c. Chafed or cracked insulation on cables.
     - d. Ignition coil weak, or inoperative.
     - e. Moisture on ignition coil, terminals, distributor cover, spark plug porcelains, or in distributor.
     - f. Cracked distributor cap.
     - g. Improper installation of secondary cables (not correct for engine firing order).
     - h. Spark plugs damaged, dirty, wet, procelains cracked, or gap improperly spaced.
     - i. Rotor contact spring bent or broken.
     - j. Distributor rotor damaged or grounded.
     - k. Distributor cap center terminal (inner) broken or missing.
  3. Ignition switch.
     - a. Loose contacts.
     - b. Corroded or burned contacts.
     - c. No by-pass circuit.
- C. Starter motor.
  1. Inoperative or not operating properly.
  2. Congealed engine oil due to the use of too heavy oil or the formation of sludge.
  3. Water in cylinders causing hydrostatic lock.
  4. Starter motor pinion stuck in flywheel gear.
  5. Clutch slipping in drive pinion.
  6. Starter switch not operating properly.

7. Faulty neutral safety switch on cars with automatic transmissions.
8. Defective starter solenoid.

3. Poor engine compression resulting from:
   A. Loose cylinder head cap screws.
   B. Spark plugs loose in head.
   C. Improperly installed or damaged cylinder head gasket.
   D. Poorly seating valves.
   E. Weak or broken valve springs.
   F. Valves holding open due to insufficient valve clearance.
   G. Valves holding open due to stem being warped, corroded, or gummed.
   H. Badly worn, broken, weak, or stuck piston rings.
   I. Hole in piston.
   J. Scored cylinder walls.
4. Unusual causes.
   A. Valves improperly timed.
   B. Camshaft timing gear broken or timing chain jumped.
   C. Cracked cylinder block.
   D. Excessive internal friction of engine assembly.

SPARK PLUG PROBLEMS

1. Plug does not fire or spark is weak.
   A. Porcelain cracked.
   B. Internal porcelain carbonized or burned.
   C. Moisture or dirt accumulation on external porcelain.
   D. Electrode gap not properly spaced.
   E. Weak ignition coil.
   F. Defective spark cables.
2. Electrodes and porcelain burn at low milages.
   A. Plug heat range too high.
   B. Use of certain type of gasoline having detrimental effect on porcelain.
   C. Excessively lean carburetor mixture.
   D. High cylinder combustion temperature.
   E. Advanced ignition timing.
3. Fouled plugs.
   A. Plugs heat range too cold.
   B. Excessively rich carburetor mixture.
   C. Engine oil passing piston rings.
4. Use of some types of spark plugs and coil suppressors in radio installations.
5. Incorrect polarity of voltage at spark plug.

ENGINE MISFIRES WHEN IDLING— NORMAL ENGINE TEMPERATURES

1. Improper carburetion resulting from:
   A. Float level too high.
   B. Float level too low.
   C. Incorrect or loose jets.
   E. Restricted or partially clogged idle air passage or jet.
   E. Choke system not operating properly.
   F. Air leak occuring between upper and lower carburetor body idle tube.
   G. Air leak occuring around the carburetor throttle shaft.
   H. Idle mixture not adjusted properly.
2. Air leaks in intake manifold or carburetor resulting from:
   A. Loose manifold connections or leaks occuring in vacuum lines and vacuum operated accessories.
   B. Loose manifold nuts or cap screws.
   C. Broken or damaged intake manifold or carburetor gaskets.
   D. Crack in manifold.
   E. Warped or damaged manifold contacting surface.
3. Incorrect ignition timing.
4. Weak ignition coil.
5. Spark plug malfunction.
6. Ignition wires open or leaking to ground.
7. Uneven compression.
8. Unusual causes.
   A. Slight water leaks occurring in the cylinder or combustion chamber.
   B. Air leaks occurring around the intake valve stem because of excessive valve stem-to-guide clearance.
9. Low battery voltage.
10. Defective PCV system.
11. Defective ignition condenser.
12. Spark advanced too far.
13. Defective ignition switch.
14. Defective valves.

LACK OF POWER OR HIGH SPEED PERFORMANCE.

1. Insufficient or unequal engine cylinder compression.
2. Improper ignition timing.
3. Inoperative manifold heater valve (valve held in closed position).
4. Improper carburetion.
5. Restricted carburetor air inlet resulting from:
   A. Dirty carburetor air cleaner.
   B. Choke valve not completely opening.
6. Throttle linkage not properly adjusted and carburetor throttle valve not completely opening.
7. Carburetor accelerating pump not functioning properly.
8. Improper fuel pump operation.

9. Partially restricted or clogged exhaust pipe, muffler, or tail pipe.
10. Excessive engine temperatures.
11. Preignition.
12. Excessive engine friction resulting from:
    A. Inadequate internal clearances (bearings and pistons).
    B. Use of extreme pressure piston rings.
13. Clutch slippage.
14. Excessive rolling resistance resulting from:
    A. Dragging brakes.
    B. Tight wheel, pinion, differential, or transmission bearings.
    C. Misalignment in power transmitting units.
    D. Misalignment of rear axle.
    E. Underinflated tires.
15. Incorrect rear axle gear ratio or oversize tires.
16. Incorrect valve timing.
17. Worn lobes on cam shaft.
18. Inaccurate speedometer (gives impression of lack of performance).
19. Defective spark plugs.
20. Defective ignition points.
21. Faulty ignition coil.

SPARK KNOCK (DETONATION)

Spark knock (detonation) causes a metallic ringing sound, often described as a "ping" and is usually encountered when the engine is laboring, accelerating rapidly, or overheating.

1. Large carbon deposits in combustion chamber.
2. Ignition timed too early.
3. Faulty distributor advance governor (weak springs).
4. Inoperative spark advance (vacuum).
5. Spark plugs.
    A. Incorrect type of plug (using a plug which is too hot).
    B. Porcelains or electrodes carbonized or burned.
6. Sharp metallic edges in combustion chamber.
7. Cylinder head gasket projecting into combustion chamber.
8. Hot engine valves resulting from:
    A. Incorrect width of valve seats.
    B. Insufficient valve clearance.
    C. Use of wrong type of valve.
    D. Thin-edged valves.
9. Lean fuel mixtures.
10. Excessive engine temperatures.
11. Poor grade of fuel.
12. Old or stale fuel.
13. Inoperative manifold heater valve (valve held in closed position).
14. Inoperative emission controls.

ENGINE BACKFIRING THROUGH CARBURETOR

1. Improper ignition timing.
2. Improperly seating valves, especially intake.
3. Incorrect valve timing.
4. Worn lobes on cam shaft.
5. Preignition from any source.
6. Excessively lean carburetor mixture.
7. Intake manifold air leaks.
8. Defective cylinder head gasket (especially between cylinders).
9. Poor quality of fuel.
10. Secondary wires improperly installed (crossed) in distributor cap.
11. Distributor centrifugal weights sticking.
12. Engine cold and choke too lean.
13. No accelerator pump action.

ENGINE CONTINUES TO RUN AFTER IGNITION IS TURNED OFF (DIESELING)

1. Idle speed set too high.
2. Improper engine timing.
3. Too lean a fuel mixture.
4. Too high a heat range on spark plugs.
5. Defective idle stop solenoid (Controlled Combustion System).
6. Inoperative emissions controls.
7. Excessive engine temperature.

COOLING SYSTEM DIAGNOSIS EXCESSIVE ENGINE TEMPERATURES

1. Ignition timing too late or too early.
2. Engine fan belt slipping.
3. Defective radiator cap.
4. Abnormal water loss from cooling system.
5. Radiator tubes restricted or clogged.
6. Radiator core surface restricted by grille, covers, emblems, heavy paint, insects, or accumulation.
7. Engine thermostat not opening properly.
8. Engine thermostat reversed when installed.
9. Deteriorated or collapsed water hose.
10. Water pump impeller loose on shaft.
11. Abnormal clearance of impeller in water pump housing.
12. Defective water pump or pump seal.
13. Abnormal sludge or dirt accumulation in radiator or water jacket of engine block.
14. Any condition that will result in preignition.
15. Spark advance stuck and not advancing timing.
16. Spark advance stuck in advanced position, not retarding at proper RPM.
17. Restriction of water transfer holes in engine block or cylinder head.

18. Incorrect cylinder head gasket or improperly installed gasket.
19. Engine fan blades bent.
20. Foreign matter in cylinder head, which obstructs water circulation.
21. Clogged exhaust system.
22. Defective thermostatic spark advance control valve.
23. A malfunctioning emission system.
24. High frictional resistance in engine assembly resulting from:
    A. Insufficient internal clearance.
    B. Internal misalignment.
    C. Insufficient oil circulation.
25. Dragging brakes.
26. Seized wheel bearings.
27. Abnormal frictional resistance in power transmitting units.
28. Overload on engine—towing trailer.
29. Use of certain types of anti-freeze solutions in warm weather.
30. Clutch slipping.
31. Radiator cap defective or wrong type.
32. Defective fan drive clutch (variable speed fans).

WATER LOSS FROM COOLING SYSTEM

1. Defective radiator cap.
2. Radiator leaks.
3. Radiator or water pump hose leakage.
4. Cooling system drain plug or petcock leakage.
5. Water pump leakage.
6. Cooling system gasket leakage.
7. Cylinder block or cylinder head cracked (leaking externally or internally).
8. Combustion gases leaking into cooling system because of poor seal at cylinder head gasket.
9. Engine overheating resulting in water boiling and loss through overflow pipe.
10. Loose cylinder head bolts.

Appendix 2

Tune-up Specifications References

Whenever tune-up work is done on an engine it is usually necessary to know its tune-up specifications. This information can be found in the repair manual for the vehicle under test. In addition, manufacturer's specifications can also be found in the following publications:

Automotive Electric Association
Suite 202 - Executive Plaza Bldg.
1301 West 22nd Street
Oak Brook, Illinois 60521

Chilton's Auto Repair Manual
Chilton Company
56th and Chestnut Streets
Philadelphia, Pa. 19139

Motors' Auto Repair Manual
250 West 55th Street
New York, New York 10019

National Service Data Book
National Automotive Service, Inc.
Div. of Glenn Mitchell Manuals, Inc.
P.O. Box 10465
San Diego, California 92110

Appendix 3

Directory of Test Equipment Manufacturers

The following is a list of automotive test equipment manufacturers.

AC Spark Plug Division
1300 N. Dort Highway
Flint, Michigan 48556

Allen Electric & Equipment Co.
2101 N. Pitcher
Kalamazoo, Michigan 49007

Amtron Precision Equipment
2501 Cork Street
Kalamazoo, Michigan 49001

Arrow Brand
80 E. Jackson Blvd.
Chicago, Illinois 60604

Associated Equipment Corporation
1541 Salzman Avenue
St. Louis, Missouri 63133

Auto-Marine Instrument Corporation
7555 N. Linder
Skokie, Illinois 60076

Autoscan Incorporated
3641 Holdrege Avenue
Los Angeles, California 90016

Borg-Warner Corporation
Automotive Parts Division
11045 Gage Avenue
Franklin Park, Illinois 60131

C. E. Niehoff and Company
4925 W. Lawrence Avenue
Chicago, Illinois 60630

Christie Electric Corporation
3410 West 76th Street
Los Angeles, California 90043

Clayton Manufacturing Company
El Monte, California 91734

Dixon, Inc.
P.O. Box 1449
Grand Junction, Colorado 81501

Fox Products Co.
4720 N. 18th Street
Philadelphia, Pennsylvania 19141

Fox Valley Instrument Company
Route 5, Box 390
Cheboygan, Michigan 49721

Hanson-Hawk, Inc.
20327 Nordhoff Street
Chatsworth, California 91311

Heyer Industries, Inc.
744 Broad
Newark, New Jersey 07102

Indy Automotive
2334 Production Drive
Indianapolis, Indiana 46241

Kal-Equip Company
P.O. Box 188
Otsego, Michigan 49078

King Electronics & Manufacturing Corp.
9123 Inman Avenue
Cleveland, Ohio 44105

Lincoln St. Louis
4010 Gordfellow Blvd.
St. Louis, Missouri 63120

Marquette Corporation
5075 Wayzata Blvd.
Minneapolis, Minnesota 55416

Milton Manufacturing Co., Inc.
1900 W. Hubbard Street
Chicago, Illinois 60622

Monit Automotive Products
4535 W. Fullerton Avenue
Chicago, Illinois 60639

Olson-Horiba
24525 Southfield Road
Southfield, Michigan 48075

Peerless Instrument Company
3910 W. Devon
Chicago, Illinois 60659

Proto Professional Equipment Co.
2209 Santa Fe Avenue
Los Angeles, California 90054

Rite Autotronics Corporation
3485 S. La Cienega Blvd.
Los Angeles, California 90016

Schauer Manufacturing Corp.
4500 Alpine Avenue
Cincinnati, Ohio 45230

Simpson Electric Company
5200 W. Kinzie Street
Chicago, Illinois 60644

Stewart-Warner
Instrument Division
1840 Diversey Parkway
Chicago, Illinois 60614

Sun Electric Corporation
Harlem and Avondale
Chicago, Illinois 60631

Triple - A Specialty Co.
5750 W. 51st Street
Chicago, Illinois 60638

Appendix 4

International Auto Tune-up Symbols

The following list consists of the symbols that are used by manufacturers of automotive test equipment on meter dials or instrument panels. Until recently the use of symbols has been more-or-less confined to Europe. Now it appears that more and more American manufacturers are beginning to incorporate them on equipment intended for use in this country.

SYMBOLS

Angle

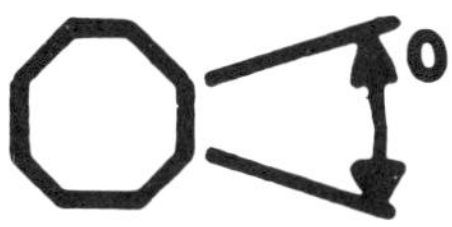

Dwell Angle

Spark

Coil Output

Points

Point Resistance

Revolution Per Minute (RPM)

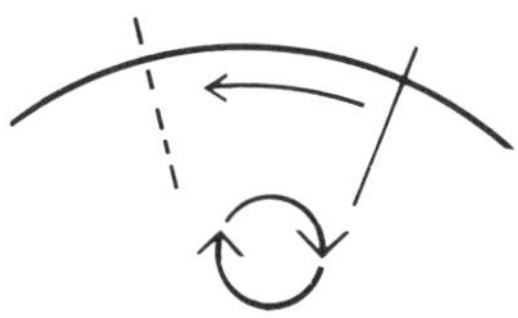

Drop in RPM

Set

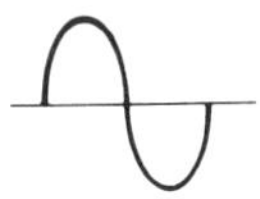

Alternating Signal

Alternator Output

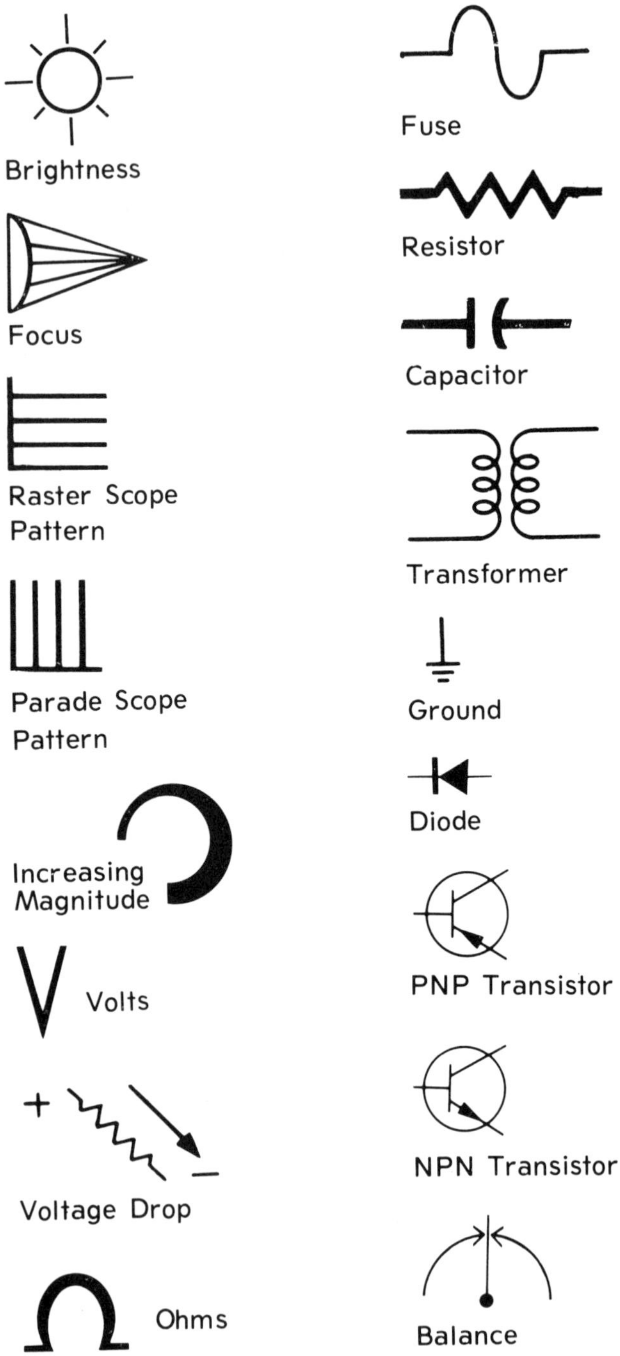
Brightness
Focus
Raster Scope Pattern
Parade Scope Pattern
Increasing Magnitude
Volts
+
−
Voltage Drop
Ohms
Fuse
Resistor
Capacitor
Transformer
Ground
Diode
PNP Transistor
NPN Transistor
Balance

Glossary

Automotive Analyzer - An instrument containing numerous automotive test features combined into one portable unit.

AC - Alternating current.

Accelerator Pump - A small pump, located in the carburetor, that sprays additional gasoline into the air stream during acceleration.

Acid - Term applied to the electrolyte in the battery, which is a mixture of sulfuric acid and water.

Advance - The moving ahead of the ignition spark; this is produced by centrifugal or vacuum devices in accordance with engine speed and intake-manifold vacuum.

Alternator - An AC generator; a generator that produces alternating current.

Alternating Current - An electric current that first flows one way in the circuit and then the other. This is the type used in homes.

Ammeter - An electric meter for measuring the amount, or number of amperes, of current flowing in an electric circuit.

Amperage - The amount of current flowing, in amperes.

Ampere - A unit of electric current.

Atmospheric Pressure - Pressure exerted by the atmosphere on all things exposed to it. Although it varies constantly, it is considered as a standard to be normal when it is 14.7 pounds per square inch at sea level.

Battery Post Adapter - A device connnected to a battery post and used for connecting ammeter leads and simulating a charged battery.

Ballast Resistor - A resistor placed in the circuit between the ignition switch and the coil to reduce the battery voltage to the coil.

Battery - A group of electric cells connected together.

Battery Charging - The process of renewing the battery by passing an electric current through the battery in a reverse direction.

B.D.C. - Bottom dead center (during an engine piston stroke).

Bendix Drive - A type of cranking motor drive which meshes with the flywheel teeth for cranking. It uncouples automatically as engine speed increases.

Blowby - The escape of exhaust gases past the piston rings.

Bourdon Tube - A curved, hollow piece of metal that is used in some instruments. Pressure inside the hollow section causes it to attempt to straighten. Vacuum inside the hollow section causes it to curl tighter. The free end then moves a needle on the gauge face.

Brush - A block of conducting substance, such as carbon, which rests against a rotating ring or commutator to form a continuous electric circuit.

Breaker Arm - The moveable arm upon which one of the breaker points is affixed.

Breaker Points - (Ignition) A pair of movable points that are opened and closed to break and make the primary circuit.

Breather Pipe - A pipe opening into the interior of the engine. It is used to assist ventilation. The pipe usually extends downward to a point just below the engine so that the passing air stream will form a partial vacuum thus assisting in venting the engine.

B.T.U. - British Thermal Unit. A measurement of the amount of heat required to raise the temperature of 1 pound of water, 1 degree Fahrenheit.

Bypass Valve - A valve that can open and allow a fluid to pass through in other than its normal channel.

Cables - Stranded conductors, usually covered with insulating material, used for connections between electrical devices.

Cam - The ignition-distributor breaker cam, which, in rotating, causes contact points to open and close.

Cam Angle or Dwell (Ignition) The number of degrees the breaker cam rotates from the time the breaker points close until they open again.

Centigrade - A temperature measuring scale. On the centigrade scale, the freezing point of water (32 degrees Fahrenheit) occurs at 0 degrees.

Carbon Pile - A pile, or stack, of carbon disks enclosed in an insulating tube. When the disks are pressed together, the resistance of the pile is decreased.

Carburetor - The device in the fuel system which mixes air and gasoline (vaporizing the gasoline as it does so) in varying proportions to suit engine operating conditions.

Cell - An electric device consisting of plates of two different chemicals surrounded by a third chemical substance. The cell produces electric voltage and current.

Centrifugal Advance - (Ignition) Spark advance resulting from engine speed.

Charging Rate - The number of amperes flowing from the charging system into the battery.

Choke - A butterfly valve, located in the carburetor, that is used to enrich the mixture for starting the engine when cold.

Circuit - (Electrical) A path for the flow of electricity from the source through the control and utilizing units and back to the source.

Coil - See Ignition Coil.

Combustion - The process involved during burning.

Combustion Chamber - The space within a cylinder above the piston. The head of the piston, the cylinder wall and the engine head form the cylinder.

Commutator - A series of copper bars that are connected to the armature windings. The bars are insulated from each other and from the armature. The brushes (as in the generator or starter) rub against the whirling commutator.

Compression Check - A test of the compression in all the cylinders at cranking speed. All plugs are removed, the compression gauge placed in one plug hole, the throttle cranked wide open and the engine cranked until the gauge no longer climbs. The compression check is a good way to determine the condition of the valves, rings and cylinders.

Compression Gauge - A gauge used to test the compression in the cylinders.

Compression Ratio - Relationship between the cylinder volume (clearance volume) when the piston is on T.D.C. and the cylinder volume when the piston is on B.D.C.

Condenser - Also called capacitor. A device in the ignition circuit which is connected across the contact points to reduce arcing by providing a storage place for electricity as points open.

Conductor - Any metallic device that will carry current.

Console - An instrument combining many automotive testers (usually including a scope) built into one large unit.

Cranking Motor - (Starter motor) The electric motor that cranks the engine, or turns the crankshaft, for starting.

Crankshaft - The main rotating memeber, or shaft, of the engine, with cranks to which the connecting rods are attached.

Current - A flow of electronics: electric current.

Current-Limiter Relay - A relay that opens its contacts when the current flowing in a circuit exceeds a certain preset value.

Current Regulator - A regulating device that limits generator output and prevents excessive generator output by repeatedly inserting a resistance into the generator field circuit.

Cutout - (Exhaust) A form of bypass valve, located in the exhaust line, that can be used to divert the flow of exhaust from one pipe to another. Often used to bypass the muffler into a straight pipe.

Cutout Relay - A device in the circuit between the generator and battery which closes for the generator to charge the battery and opens when the generator stops.

Cylinder - A tubular-shaped structure. In the engine, the tubular housing in which the piston moves up and down.

Cylinder Head - The metal section that is bolted on top of the engine block. It is used to cover the tops of the cylinders. In many cases the cylinder head contains the valves. It also forms part of the combustion chamber.

DC - Direct Current.

Diaphragm - A flexible membrane clamped at the edges and usually spring loaded; used in various pumps and controls.

Direct Current - Electrical current that flows in one direction only, such as the current supplied by a battery.

Discharge - (Battery) The action of drawing electric current from a battery, or the result of such an action.

Distributor - (Ignition) A unit designed to make and break the ignition primary circuit and to distribute the resultant high voltage to the proper cylinder at the correct time.

Distributor Cap - (Ignition) An insulated cap containing a central terminal with a series (one per cylinder) of terminals that are evenly spaced in a circular pattern around the central terminal. The secondary voltage travels to the central terminal where it is then channeled to one of the outer terminals by the rotor.

Dry Charged Battery - A battery with the plates charged but lacking electrolyte. When ready to be placed in service, the electrolyte is added.

Dual Breaker Points - (Ignition) A distributor using two sets of breaker points to increase the cam angle so that at high engine speeds sufficient spark will be produced to fire the plugs.

ECS - Exhaust Control System.

Electro-Chemical - Chemical (battery) production of electricity.

Electrochemical Device - A device that operates on both electrical and chemical principles (a lead-acid storage battery, for example).

Electrode - (Spark plug) The center rod passing through the insulator forms one electrode. The rod welded to the shell forms another. They are referred to as the center and side electrodes.

Electromagnet - A coil of wire (usually with an iron core) which produces magnetism when electric current passes through it.

Energy - The ability to do work.

Engine Displacement - The volume of the space through which the head of one piston moves in the full length of its stroke, multiplied by the number of cylinders in the engine. The result is given in cubic inches.

Exhaust Gas Analyzer - An instrument used to check the exhaust gases to determine combustion efficiency.

Field Coil - A coil, or winding, in a generator or cranking motor which produces a magnetic field as current passes through it.

Firing Order - The order in which cylinders must be fired - 1, 5, 3, 6, 2, 4 etc.

Float Bowl - That part of a carburetor that acts as a reservoir for gasoline and in which the float is placed.

Fuel Injection - A fuel system that uses no carburetor but sprays fuel either directly into the cylinders or into the intake manifold just ahead of the cylinders.

Fuel Mixture - A mixture of gasoline and air.

Fuel Pump - A vacuum device, operated either mechanically or electrically, that is used to draw gasoline from the tank and force it into the carburetor.

Fuel System - In the automobile, the system that delivers to the engine cylinders the combustible mixture of vaporized fuel and air. It consists of fuel tank, lines, gauge, pump, carburetor, and manifold.

Fuse - A protective device that will break the flow of current when current draw exceeds the capacity of the fuse.

Galvanometer - An instrument used to measure the presence, amount of, and direction of an electric current.

Gasoline - A hydrocarbon fuel used in the internal combustion engine.

Generator - That part of the electric system which converts mechanical energy into electric energy for lighting lights, charging the battery, operating the ignition system, and so on.

Hydrocarbon - A mixture of hydrogen and carbon.

Hydrometer - A device used to measure specific gravity; especially, the hydrometer used to measure the specific gravity of battery electrolyte for an indication of state of battery charge. Another type of hydrometer is used to measure anti-freeze in the radiator.

Idle Valve or Idle Needle - A needle used to control the amount of fuel mixture reaching the cylinders during idling. It, or they, may be adjusted by turning the exposed heads.

Ignition Coil - That part of the ignition system which acts as a transformer to step up the battery voltage to many thousands of volts. The high-voltage surge then produces a spark at the spark-plug gap.

Ignition System - In the automobile, the system that furnishes high-voltage sparks to the engine cylinders to fire the compressed air-fuel charges. It consists of battery, ignition coil, ignition distributor, ignition switch, wiring, and spark plugs.

Insulator - A device made of a material that strongly opposes flow of electric current through it.

Intake Manifold - Connecting tubes between the base of the carburetor and the port openings to the intake valves.

Internal Combustion Engine - An engine that burns fuel within itself as a means of developing power.

Jet - A small hole or orifice used to control the flow of gasoline in various parts of the carburetor.

LPG - Liquified Petroleum Gas.

Magneto - An engine-driven unit that generates high voltage to fire the spark plugs. It needs no outside source of power such as a battery.

Manifold Heat Control Valve - A valve placed in the exhaust manifold, or in the exhaust pipe, that deflects a certain amount of hot gas around the base of the carburetor to aid in warmup.

Ohm - A unit of electric resistance.

Ohmmeter - An instrument used to measure the amount of resistance in a given item or circuit.

Overcharging - Continued charging of a battery after it has reached a charged condition. This action damages the battery and shortens its life.

Parallel Circuit - A circuit in which all negative terminals are connected together and all positive terminals are connected together. One electrical device can be operated independently of all others in a parallel circuit.

PCV - Positive Crankcase Ventilation.

Primary Circuit - (Ignition system) The low voltage (6 to 12) part of the ignition system.

PSI - Pounds per square inch.

Recharging - The action that takes place when electric current is forced into a battery in a direction opposite to that in which current flows from the battery when the battery is being discharged.

Rectifier - A device to change alternating current to direct current.

Regulator - In the electric system, a device that controls the charging system output to prevent excessive voltage or excessive output.

Resistance - That property of a material that opposes the flow of electric current.

Resistor - A device placed in a circuit to lower the voltage. It will also decrease the flow of current.

Resistor Spark Plug - A spark plug containing a resistor, designed to shorten both the capacitive and inductive phases of the spark. That will suppress radio interference and lengthen electrode life.

Retard - (Ignition timing) To set the ignition timing so that a spark occurs later or less degrees before T.D.C.

RPM - Revolutions Per Minute.

Secondary Circuit - (Ignition system) The high voltage part of the ignition system.

Self Discharge - Chemical activities in the battery which cause the battery to discharge even though it is furnishing no current.

Separator - The thin sheet of wood, rubber, or glass mat that is placed between positive and negative plates in a battery cell to insulate them from each other.

Series Circuit - An electric circuit in which the same current flows through all devices; positive terminals are connected to negative terminals.

Short - A short circuit; a defect in an electric circuit which permits current to take a short path, or circuit, instead of following the prescribed path.

Shunt - A parallel connection, or circuit.

Slip Rings - In an AC machine, the rings which form rotating connections between the armature windings and the brushes.

Solenoid - A coil of wire: As used in the cranking-motor circuit, for example, a coil of wire with a movable metal core. Energizing the coil (by connecting it to the battery) pulls the core in, causing the cranking-motor drive pinion to mesh and the cranking-motor contacts to close.

Solenoid Switch - A switch which is opened and closed electromagnetically, by movement of a core. Usually, this core also causes a mechanical action, such as the movement of a drive pinion into mesh with flywheel teeth for cranking.

Spark - The bridging or jumping of a gap between two electrodes by a current of electricity.

Spark Advance - Causing the spark plug to fire earlier by altering the position of the distributor breaker points in relation to the distributor shaft.

Spark Plug - An assembly which includes a pair of electrodes and insulator. It provides a spark gap in the engine cylinder.

Specific Gravity - A measure of the weight per unit volume of a liquid as compared to the weight of an equal volume of water.

Starting Motor - Same as cranking motor.

Storage Battery - That part of the electric system which acts as a reservoir for electric energy, storing it in chemical form.

Sulfation - The lead sulfate that forms on battery plates as a result of the battery action that produces electric current.

Switch - A device that opens and closes an electric circuit.

Tachometer - A device used to indicate the speed of the engine in rpm.

T.D.C. - Top Dead Center.

Throttle Valve - A valve in the carburetor. It is used to control the amount of fuel mixture that reaches the cylinders.

Timing - The procedure of adjusting the ignition distributor so the sparks will occur at the spark plugs at the correct time in relation to the position of the piston on the compression stroke.

Timing Chain - A drive chain that operates the camshaft by engaging sprockets on the cam shaft and crankshaft.

Timing Gears - Both the gear attached to the camshaft and the gear on the crankshaft. They provide a means of driving the camshaft.

Timing Marks - (Ignition) Marks, usually located on the vibration damper, used to synchronize the ignition system so that the plugs will fire at the precise time.

Timing Light - A stroboscope unit that is connected to the secondary ignition circuit to produce flashes of light in unison with the firing of a specific spark plug. By directing these flashes of light on the whirling timing marks, the marks appear to stand still. By adjusting the distributor, the timing marks may be properly aligned thus setting the timing.

Troubleshooting - The work necessary to discover the cause of a trouble; also implies the correction of the trouble by elimination of the cause.

Tune-Up - The process of checking, repairing, and adjusting the carburetor, spark plugs, belts, points, timing, etc., in order to obtain the maximum performance from the engine.

Tune-Up Kit - Individual instruments packaged and sold together as one basic kit for performing automotive tune-up work.

Twenty-Hour Rate - A battery rating; amount of current a battery can deliver for 20 hours without cell voltage dropping below 1.75 volts, starting with an electrolyte temperature of 80 degrees F.

Vacuum - An enclosed area in which the air pressure is below that of the surrounding atmospheric pressure.

Vacuum Advance - Ignition-spark advance resulting from partial vacuum in intake manifold.

Vacuum Pump - A diaphragm type of pump used to produce a vacuum.

Volt - A unit of electrical pressure or force that will move a current of one ampere through a resistance of one ohm.

Voltage - A difference in electrical potential between one end of a circuit and the other. Also called EMF (electromotive force). Voltage causes current to flow.

Voltage Drop - The drop, or reduction, in voltage across a resistance (a circuit, connection, electrical device, etc.).

Voltage Regulator - A regulating device that prevents excessive charging system voltage by repeatedly inserting a resistance into the generator-field circuit.

Voltmeter - An electric meter for measuring the voltage of an electric device, such as a battery or generator, or for measuring the voltage between two points in an electric circuit.

ILLUSTRATION CREDITS

The illustrations listed below have been supplied by the courtesy of, and with the permission of, the companies shown.

PEERLESS INSTRUMENT CO. for the cover photograph and for Figures: 1-1; 1-3 through 1-6; 1-8; 1-11; 1-13 through 1-20; 2-6; 2-10 through 2-13; 2-15; 3-1 through 3-2; 3-4; 3-6 through 3-7; 3-9; 3-11 through 3-20; 3-22 through 3-25; 4-1 through 4-2; 4-4 through 4-20; 4-22; 5-1 through 5-5; 5-9 through 5-13; 6-8 through 6-10; 7-1; 7-9 through 7-13; 8-3; 8-5 through 8-6; 8-13 through 8-14; 8-16 through 8-18; 8-21 through 8-38; 8-40 through 8-41; 9-1; 9-5; 9-7 through 9-9; 10-1; 10-3 through 10-12; 11-1; 11-3; 11-5; 11-8 through 11-11; 12-9 through 12-12; 12-17.

SUN ELECTRIC CO. for Figures: 1-7; 1-9; 2-1; 2-4; 2-7; 2-8; 2-14; 3-5; 3-8; 3-10; 3-21; 4-3; 4-25; 7-4; 6-11; 8-1; 8-4; 8-7; 8-8; 8-10 through 8-12; 8-15; 8-39; 9-2 through 9-4; 10-13; 11-4; 11-6; 11-7; 11-12 through 11-14; 12-1 through 12-5; 12-8; 12-14; 12-15.

KAL EQUIP COMPANY for Figures: 3-3 and 7-2.

MARQUETTE CORPORATION for Figures: 3-27; 11-15; 11-16.

FOX VALLEY INSTRUMENT COMPANY for Figures: 4-21; 5-7; 6-6.

PIONEER ELECTRIC & RESEARCH CO. for Figure: 4-23.

ALLEN ELECTRIC AND EQUIPMENT CO. for Figure: 6-3.

SIMPSON ELECTRIC CO. for Figure: 6-5.

KING ELECTRONICS for Figure: 8-2.

AUTOSCAN INC. for Figure: 12-13.

AC SPARK PLUG DIVISION for Figure: 12-16.

Index

V